T0237563

SpringerBriefs in Applied Sciences and Technology

Computational Intelligence

Series Editor

Janusz Kacprzyk, Systems Research Institute, Polish Academy of Sciences, Warsaw, Poland

SpringerBriefs in Computational Intelligence are a series of slim high-quality publications encompassing the entire spectrum of Computational Intelligence. Featuring compact volumes of 50 to 125 pages (approximately 20,000–45,000 words), Briefs are shorter than a conventional book but longer than a journal article. Thus Briefs serve as timely, concise tools for students, researchers, and professionals.

More information about this subseries at http://www.springer.com/series/10618

Surbhi Bhatia · Poonam Chaudhary ·
Nilanjan Dey

Opinion Mining
in Information Retrieval

 Springer

Surbhi Bhatia
Department of Information Systems
King Faisal University
Al Hasa, Saudi Arabia

Nilanjan Dey
Department of Information Technology
Techno India College of Technology
Kolkata, India

Poonam Chaudhary
Department of Computer Science
and Engineering and Information
Technology
The NorthCap University
Gurugram, India

ISSN 2191-530X   ISSN 2191-5318   (electronic)
SpringerBriefs in Applied Sciences and Technology
ISSN 2625-3704   ISSN 2625-3712   (electronic)
SpringerBriefs in Computational Intelligence
ISBN 978-981-15-5042-3   ISBN 978-981-15-5043-0   (eBook)
https://doi.org/10.1007/978-981-15-5043-0

© The Author(s), under exclusive license to Springer Nature Singapore Pte Ltd. 2020
This work is subject to copyright. All rights are solely and exclusively licensed by the Publisher, whether the whole or part of the material is concerned, specifically the rights of translation, reprinting, reuse of illustrations, recitation, broadcasting, reproduction on microfilms or in any other physical way, and transmission or information storage and retrieval, electronic adaptation, computer software, or by similar or dissimilar methodology now known or hereafter developed.
The use of general descriptive names, registered names, trademarks, service marks, etc. in this publication does not imply, even in the absence of a specific statement, that such names are exempt from the relevant protective laws and regulations and therefore free for general use.
The publisher, the authors and the editors are safe to assume that the advice and information in this book are believed to be true and accurate at the date of publication. Neither the publisher nor the authors or the editors give a warranty, express or implied, with respect to the material contained herein or for any errors or omissions that may have been made. The publisher remains neutral with regard to jurisdictional claims in published maps and institutional affiliations.

This Springer imprint is published by the registered company Springer Nature Singapore Pte Ltd.
The registered company address is: 152 Beach Road, #21-01/04 Gateway East, Singapore 189721, Singapore

Preface

The core part of human actions that play a vital role on the conduct and behavior is the 'opinions.' Opinions can be considered as a feeling, review, sentiment, or assessment of an object, product, or entity. The way of expressing opinions on certain products that people purchase and the services that they receive in the various industries has been transformed considerably because of ubiquitous webbing. So, people have been inclined to engage themselves more in online shopping. Review sites and social networking sites fascinate people to post feedbacks and reviews online on blogs, Internet forums, review portals, and on many more platforms. These opinions play a very important role for customers and product manufacturers as they tend to give better knowledge of buying and selling by setting positive and negative comments on products and other information which can improve their decision-making policies.

Mining of such opinions has focused the researchers to pay a keen intention in developing such a system which not only collects useful and relevant reviews online in a ranked manner but also produces an effective summary of such reviews collected on different products according to their respective domains. However, there is little evidence that researchers have approached this issue in opinion mining with the intent of developing a system. Existing opinion mining systems lacked several important features such as aspect detection, summaries generated on the basis of aspects, and classification of opinions using emerging learning algorithms. Existing research in sentiment analysis tends to focus on finding out how to classify the opinions using traditional machine learning algorithms and produce a collaborative summary in their respective domains. In spite of an increase in the field of opinion mining and its research, presenting all the areas such as aspect identification, opinion classification, and opinion summarization together by developing a coherent structure as an information retrieval system is still lacking.

This book intends to design a more comprehensive way of building a system to mine opinions and present an integrative framework with hybrid techniques of learning. Consequently, the aim of this book is to discuss the overall novel architecture of developing an opinion system that will address the remaining challenges and provide an overview of how to mine opinions. The people's opinions which are extracted from the writings on the Web largely depend on the above-mentioned factors, and the summary formed should be structured and concise in the aggregated form covering the important aspects of the product.

Opinion mining is an integral part of many commercial applications and research projects today, in areas ranging from data mining, information retrieval, and machine learning to finding your friends on social networks. In this book, we want to show you how easy it can be to build information retrieval system and how to best go about it. The applications of opinion mining are endless and, with the amount of data available today, mostly limited by your imagination.

I dedicate my work to my beloved husband Mr. Sahil and my daughter Inaaya for encouraging me to pursue my research work. They have truly been there for me to give me all the support that I needed. My special thanks to my friends and colleagues for helping me in writing this research oriented book.

Al Hasa, Saudi Arabia   Surbhi Bhatia
Gurugram, India   Poonam Chaudhary
Kolkata, India   Nilanjan Dey

About This Book

The core part of human actions that play a vital role on the conduct and behavior is the "Opinions." Opinions can be considered as a feeling, review, sentiment, or assessment of an object, product, or entity. The way of expressing opinions on certain products that people purchase and the services that they receive in the various industries has been transformed considerably because of ubiquitous webbing. So, people have been inclined to engage themselves more in online shopping. Review sites and social networking sites fascinate people to post feedbacks and reviews online on blogs, Internet forums, review portals, and on many more platforms. These opinions play a very important role for customers and product manufacturers as they tend to give better knowledge of buying and selling by setting positive and negative comments on products and other information which can improve their decision-making policies.

Mining of such opinions has focused the researchers to pay a keen intention in developing such a system which not only collects useful and relevant reviews online in a ranked manner but also produces an effective summary of such reviews collected on different products according to their respective domains. However, there is little evidence that researchers have approached this issue in opinion mining with the intent of developing a system. Existing opinion mining systems lacked several important features such as aspect detection, summaries generated on the basis of aspects, and classification of opinions using emerging learning algorithms. Existing research in sentiment analysis tends to focus on finding out how to classify the opinions using traditional machine learning algorithms and produce a collaborative summary in their respective domains. In spite of an increase in the field of opinion mining and its research, presenting all the areas such as aspect identification, opinion classification, and opinion summarization together by developing a coherent structure as an information retrieval system is still lacking.

This book intends to design a more comprehensive way of building a system to mine opinions and present an integrative framework with hybrid techniques of learning. Consequently, the aim of this book is to discuss the overall novel architecture of developing an opinion system that will address the remaining challenges and provide an overview of how to mine opinions. The people's opinions which are extracted from the writings on the Web largely depend on the above-mentioned factors, and the summary formed should be structured and concise in the aggregated form covering the important aspects of the product. A complete dynamic information retrieval system is proposed giving summarized opinions according to the interest of the user based on the features. The fresh reviews are crawled using World Wide Web by proposing novel algorithms. The revisit frequency of the updated Web pages is computed, and the most recent reviews are extracted by opinion retriever. The pre-processing tasks are carried on the opinions for cleaning and removing irrelevant content. After the extraction is done, the following activities are carried out. Firstly, the features from the customer's opinions are identified by using dependency rules of natural language processing. Secondly, the orientation of each review based on the extracted feature is detected using deep learning algorithms. The proposed model of convolutional neural networks is used for binary opinion classification. Finally, abstractive and extractive techniques are used for summarizing the opinions by proposing novel algorithms. Comparative experiments on different datasets (standard and extracted) are conducted, and the accuracy is effectively measured using ROUGE tool.

The prime component that can enhance the quality of services is the opinion of the users. The two vital roles played by humans are sentiments and emotions which they share with their friends, relatives and associations. The ease of Internet access, economical computing devices leads the users to online shopping which has exponentially increased the data about the people's mood and opinion. This makes 'Opinion' as prime component that can enhance the quality of services. The new resources and interactive format of feedback system adopted by review sites and social media has opened huge heterogeneous data sources of user opinions. Researchers found that opinions of people are scattered on various online shopping sites and social media, so instead of traversing the different websites for the reviews with their polarity education of sentences, a complete IR system should be developed which directly gives the summarized opinions.

Opinion Mining in Information Retrieval makes available approaches and techniques for complete opinion oriented Information Retrieval systems and discusses in detail the trends of sentiment analysis emphasizing on "How online review and feedback reflects the opinion of users and lead to the major drift in the decision making process of the organization". The book focuses on providing different machine learning and deep learning approaches to solve the new challenges raised by opinion mining applications with the survey and comparison with traditional fact based studies. It includes the Opinion Score Mining System, Opinion Retrieval, Aspect Extraction. Finally, it gives the application of deep learning and machine learning approaches that develop the opinion oriented Information Retrieval systems along with the benchmark datasets, discussion on available sources, opinion summarization, and future work.

Opinion Mining in Information Retrieval is the advanced comprehensive survey of this challenging yet interesting field of Opinion Mining and will be of interest professionals, researchers and individuals have interest in this area.

Contents

About the Authors

Surbhi Bhatia is an Assistant Professor in Department of Information Systems, College of Computer Sciences and Information Technology, King Faisal University, Saudi Arabia. She has rich 8 years of teaching and academic experience. She received her Ph.D. from Banasthali Vidyapith, Rajasthan in 2018. She is in the Editorial board member with Inderscience Publishers in the International Journal of Hybrid Intelligence. She has published three patents with Government of India. She has published 20 papers in reputed journals and conferences in high indexing databases. She has successfully published many book chapters indexed in major indexing databases. She has supervised research projects at UG and PG level. Her research areas include Databases, Data mining and Machine Learning.

Poonam Chaudhary has 12 years of teaching and industry experience. She is currently working as an Assistant Professor in Department of CSE & IT, The NorthCap University, Gurugram, India. She has completed her B.Tech from University of Rajasthan, Jaipur followed by M.Tech from CITM, Faridabad affiliated by MDU Rohtak, India. Currently, she is pursuing PhD from MRIIRS Faridabad, India in the field of Brain Computer Interfacing using EEG signals. She has mentored DST Sponsored NewGen IEDC four student startup projects in the field of computer science. She has more than 10 publications to her credit in various leading International and National Journals/Conferences in the various areas like Brain Computer Interfacing, Data Mining and Machine Learning, Natural Language Processing.

Nilanjan Dey, is an Assistant Professor in Department of Information Technology at Techno International New Twon (Formerly known as Techno India College of Technology), Kolkata, India. He is a visiting fellow of the University of Reading, UK. He is a Visiting Professor at Duy Tan University, Vietnam. He was an honorary Visiting Scientist at Global Biomedical Technologies Inc., CA, USA (2012-2015). He was awarded his PhD. from Jadavpur Univeristy in 2015. He is the Editor-in-Chief of International Journal of Ambient Computing and Intelligence, IGI Global. He is the Series Co-Editor of Springer Tracts in Nature-Inspired Computing, Springer Nature, Series Co-Editor of Advances in Ubiquitous Sensing Applications for Healthcare, Elsevier, Series Editor of Computational Intelligence in Engineering Problem Solving and Intelligent Signal processing and data analysis, CRC. He has authored/edited more than 50 books with Springer, Elsevier, Wiley, CRC Press and published more than 300 peer-reviewed research papers. His main research interests include Medical Imaging, Machine learning, Computer Aided Diagnosis, Data Mining etc. He is the Indian Ambassador of International Federation for Information Processing (IFIP) – Young ICT Group.

Chapter 1
Introduction to Opinion Mining

The prime component that can enhance the quality of services is the opinion of the users. Factual information and opinion information are two categories of textual information [1]. Facts are sentences which are true and can be verified, whereas opinions are sentences which hold an element of belief and cannot be verified for their truth. Researchers rely on mining of subjective part of information which includes opinions and emotions. The increasing popularity of online shopping sites, social media, and other review interaction sites have evolved millions of users to post opinions on the World Wide Web (WWW). Whenever a person is purchasing a product, one can directly collect information on the Web, instead of relying on friends and family for their views and feedbacks. During political elections, the voting decisions can be taken similarly. Market investigations can be accomplished through mining information on Internet for gathering knowledge and interest of the customers. All these factors aim to focus more inclusively on opinion mining.

1.1 Opinions: A Cognitive Source of Information

The two vital roles played by humans are sentiment and emotion which they share with their friends, relatives, and associations. 'What are we saying?' and 'How are we feeling?' are the basic questions which govern most of our decisions. In today's world, an online presence has become a trend and posting photographs, sharing reviews, activities, and experiences have become an addiction. Though, social media enable the users to create and share private contents and thoughts informally for social networking, it is also used for more formal and public sharing of thoughts, political views, and opinions of journalists, politicians, and public figures. The ease of Internet access and cheap computing devices leads the users to online shopping which has exponentially increased the data about the people's mood and opinion.

© The Author(s), under exclusive license to Springer Nature Singapore Pte Ltd. 2020
S. Bhatia et al., *Opinion Mining in Information Retrieval*,
SpringerBriefs in Computational Intelligence,
https://doi.org/10.1007/978-981-15-5043-0_1

The new resources and interactive format of feedback system adopted by review sites and social media have opened huge heterogeneous data sources of user opinions.

Researchers found that opinions of people are scattered on various online shopping sites and social media, so instead of traversing the different Web sites for the reviews with their polarity education of sentences, a complete IR system should be developed which directly gives the summarized opinions. The need of an automated system for the logical analysis of textual information will allow users to investigate a large amount of data to process within a short span of time and understand the crux of it. By using the current modern techniques of big data, AI and data mining have already helped the researchers to make an automated system. Nowadays, machine learning models have shown promising performance on classification of specific opinion documents. Many researchers have applied well-known supervised/unsupervised machine learning algorithms for classification and clustering of data for determining the sentiments of people. This book will include hybrid techniques of machine learning and NLP based applications to develop a dynamic IR system which will be achieved by conducting extensive research and real-time monitoring of review Web sites.

1.1.1 Necessity for E-commerce: A New Trend in Online Shopping

Extraction of information from texts, news articles, and social media involves mainly the use of natural language processing (NLP). In the 1990s, Message Understanding Conference (MUC) declared information extraction (IE) to be an automatic technique for developing well-defined representation of information gathered from text documents [2]. MUC greatly contributed to research in the area of information extraction. This term 'IE' helps in identifying entities and relations existing between the entities in texts for the storage in structured databases. During the course of advancement of the Web and the availability of the Internet, considerable attention has been given to Web documents as they serve as an innovative medium of representing individual experiences in the form of reviews, opinions, blogs, comments, or feedbacks as 'consumer-generated media.'

This has led to generation of interest in technologies and in research for automatically extracting, classifying, and summarizing personal opinions from Web documents and so has led to the development of systems that mine opinions. The advancements in research toward opinion mining have subsided the earlier survey-based feedback research. This has impacted the Web users who seek reviews online for making decisions on certain products of their interest because of easy availability of online reviews. If a system can be developed that can provide users with such type of summarized reviews, then the society can be benefitted in a number of ways:

- It reduces the efforts and time people spend hunting for reliable reviews.
- People can view the opinions based on the features, one is interested in.

- People can get a cumulative count of positive and negative reviews based on the features.
- It gives an overall comprehensive compilation of opinions by fusing sentiments.

Social networking has been a need of today's society. In earlier times, people tend to buy or sell products by posting advertisements, conducting surveys, focus groups, consultants, taking opinions from friends and relatives. But, now it is not restricted to someone's circle of friends or family or small surveys, but it has spread globally to online social media as blogs, posts, tweets, social networking sites, review sites, and so on. However, the transition from surveys to social media is interesting but not that easy. The business analytical reports have shown that many organizations have improved their sales, marketing, and strategy, setting up new policies and making decision based on opinion mining techniques. This book discusses in detail the trends of sentiment analysis emphasizing on 'How online review and feedback reflects the opinion of users and lead to the major drift in the decision-making process of the organization.'

1.1.2 Facts and Opinions: Types of Opinion

Facts are sentences which are true and can be verified, whereas opinions are sentences which hold an element of belief and cannot be verified for their truth. The emotions and feeling of human being could be described as events, entities, and properties; further evaluated with subjective approach, and can be considered as an opinion. The difference between both is shown in Fig. 1.1.

For example, the sentence 'The laptop has a great battery backup' is in the subjective category, while the sentence 'The laptop weights 250 g' contains an objective evaluation. Opinions are of two types: regular and comparative. A regular direct opinion has basic four components represented as:

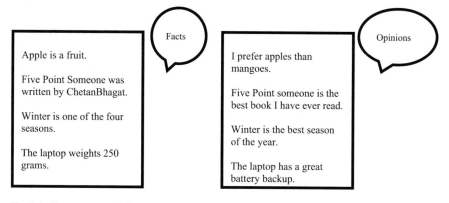

Facts

Apple is a fruit.

Five Point Someone was written by ChetanBhagat.

Winter is one of the four seasons.

The laptop weights 250 grams.

Opinions

I prefer apples than mangoes.

Five Point someone is the best book I have ever read.

Winter is the best season of the year.

The laptop has a great battery backup.

Fig. 1.1 Facts versus opinions

$$\left(\text{target}_i, \text{sentimentvalue}_i, \text{holder}_i, \text{time}_i\right) \qquad (1.1)$$

The 'target$_i$' can be described as an item, individual, society, or subject. It can be represented as a chain of components and its subparts. The target can be a single entity or a combination of many entities. 'Sentimentvalue$_i$' are emotions which reflect one's response in view of an entity categorized as positive, negative, or neutral category. The sentence as an example 'My laptop has a great interface' contains a positive evaluation, while the sentence 'This mobile phone lacks many features' contains a negative evaluation. 'Holder$_i$' is the person who holds the opinion. 'Time$_i$' is a plan or a schedule, when the opinion was posted [3, 4].

1.1.3 Demand for Information on Opinions

Researchers are focusing on enormously on individual areas under opinion mining such as aspect identification, opinion classification, and opinion summarization. But, still presenting all the areas together by developing a coherent structure as an information retrieval system is still lacking. Though efforts on developing opinion mining systems have been made previously, certain limitations are identified. This research intends to present an integrative framework as a system with hybrid techniques of learning that will address the challenges discussed above.

Research activities are on a high pace and focusing immensely on the area of opinion mining. The papers and projects discussed in the early years of 1980s [5, 6] may be seen as portent in the field. In 2003, Dave et al. [7] have analyzed the review about entities and presented a model for classification of document polarity as either recommended or not recommended. They have used the term opinion mining for the first time in the history of opinion mining, though some work had previously addressed the same task [8, 9]. This book has opened new avenues of understanding of applied machine leaning approach for research in NLP and text mining, and summarizes a few years extensive research that had been done in this area. In year 2002, Pang et al. [10] indicate the beginning of research in the subject, and in 2012, has shown a remarkable growth in the field of sentiment analysis. Several factors drive the focus on the subject. Few are mentioned as under:

- NLP and IR have become dependent on machine learning methods.
- There are large training datasets available for machine learning algorithms due to the tremendous growth of WWW.
- The increasing methodologies involving AI and other commercial applications.

1.1.4 Google: Not as Effective for Searching Opinions

Due to the ease of publishing online, social networking sites have captured the market, and they have produced tremendous growth of opinions and feedbacks commented by users over different domains online. On the applicability of innovative searching techniques of the current search engines, they serve to extract documents or Web pages with respect to particular keywords enquired. But, instead of retrieving just the documents or Web pages, the person may be concerned with retrieving opinions in summarized form mentioning its features. This is referred to as 'opinion mining.' As there is scope of improvement in the current search ranking strategy for opinion retrieval/search, this research intends to develop the system of IR covering the major aspects of opinion mining aiming to include several models, techniques, and schemes under the disciplines of computer science and including studies in databases, data mining, social media, and IR. The search will encompass various components staring from posting a user query on a particular topic, crawling opinions related to that search query to categorize opinions as positive and negative; fusing the sentiments together to present the summarized opinions based on aspects.

This book discusses all possible solutions to the problem mentioned above, but before covering the solutions, the steps in opinion mining are elaborated.

1.2 Understanding Opinion Mining

Opinion mining reclines at the crossroads of computational linguistics and IE. Opinion mining is a subfield of data mining. Data mining is referred to as extracting important information which can be represented in structured and understandable form of knowledge as graphs, patterns, etc. [11]. The authors Seerat and Azam [12] gave a different definition to opinion mining which states that automatic extraction of knowledge of opinion mining by taking opinions from different sources like online shopping sites, review sites, etc., on a specific area or problem. Opinion mining collects concepts and ideas from mainly two disciplines, i.e., information retrieval and computational linguistics (CL). The two major aspects of opinion mining are question answering (QA) and IE [13], i.e., with respect to IR. QA basically deals with answering the questions written in natural language. For example, 'What is your viewpoint regarding buying an iPhone? IE relatively performs strong subjectivity analyses or sentiment analysis using various NLP and statistical techniques. Thus, sentiment analysis is the second major term which is used by opinion mining. Considering the broader view, opinion mining and sentiment analysis go hand in hand and represent the same study field. More concrete and precise definitions of each term associated with opinion mining such as opinion holder, feature, and view are explained later in figures. The classical model of opinion mining is similar as mentioned in the book written by Liu [14].

1.2.1   Definition of Opinion Mining by Various Researchers

Opinion mining could be well-thought-out as a method of assembling reviews/feedbacks which are shared online on various social media Web sites, blogs, and e-commerce Web sites followed by the extracting information in the interior raw data. Many researchers used this definition, which is more ubiquitous. The bigger term has been reflected in this definition which comprises many subdefinitions.

Additional adopted definition states that opinion mining is identifying and abstracting materialistic and subjective facts from text documents. Review mining, sentiment analysis, appraisal extraction are some terms which can be used interchangeably by opinion mining [15]. Data mining and WWW lie in close proximity with each other. Opinion mining is enlisted as a part of Web content mining (WCM) and can be categorized in the hierarchy of Web mining [16]. Machine learning and classification can be employed at various levels of information retrieval (IR). Opinion mining is considered as a field of IR. Hence, opinion mining has a big influence of text mining techniques. According to the definition, Web mining refers to the application of text mining techniques for extracting useful knowledge from Web text. So, typically opinion mining relates Web mining too. Therefore, Web mining techniques can be hypothesized as a significant resource for opinion mining research.

Data Mining: Data mining is the practices that agreement with investigation of pre-acquired data from several perceptions and recapitulate this analyzed data into useful information that leads to data-driven decision-making companies and individuals. It can also be well-thought-out as a stride in the process of *Knowledge Discovery from data*.

Web Mining: As discussed earlier in this section, Web mining is a specific case of text mining. It uses a set of text mining techniques used on text for classification, prediction, knowledge discovery, and forecasting. It is profoundly rooted in the extraction of knowledge commonly associated with Web documents such that 'many Web search engines uses text mining techniques in order to extract the most pertinent documents as a result of a search query.' Web mining can be avowed as analyzing and extracting valuable information from WWW. Web mining is fundamentally classified as three categories: (1) Web content mining, (2) Web usage mining, and (3) Web structure mining.

Web Usage Mining (WUM): Web usage mining uses data mining techniques to extract relevant usage patterns from Web documents in order to recognize and effective way to serve the needs of Web-based applications. Text mining is the analysis of data present in natural language text. Web usage mining refers to the automatic discovery and analysis of patterns in click stream and associated data collected or generated as a result of user interactions with Web resources on one or more Web site [17].

Web Content Mining (WCM): It targets the Web documents which contain multimedia content like text, images, video, and audio for knowledge discovery. It deeply examines the contents of Web documents for relevant information extraction.

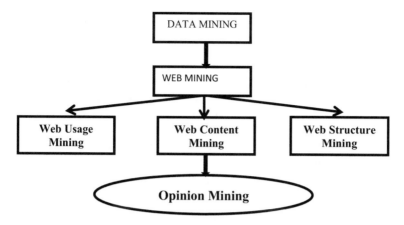

Fig. 1.2 Structure of data mining

Web Structure Mining (WSM): It generates structural summary of a Web site or a Web page by studying and analyzing the links and hyperlinks present on the same.

The hierarchical structure explained above justifies that the three terms, i.e., database, IR, and AI, are closely interrelated [18]. Figure 1.2 gives the insight in the hierarchical structure of data mining.

Determining the sentiment polarity from reviews requires a sentence-level or even aspect-level opinion mining. To find opinion of the author on each feature from the review is a classification task [12]. This task involves the use of the natural language processing techniques to make a better understanding of text for effective opinion mining. There are three prime components of an opinion mining problem, the opinion source, target entities in an opinion, and the viewpoint of the opinion holder.

Opinion mining can be redefined as a set of text documents (D) which contain opinions, features of opinion on the document $f \in D$ and classification categories, positive comments, negative or neutral comments.

This can be considered as of the most appropriate definition of opinion mining as it incorporates all components and their dependency involved with each other. The resonance of this definition with IR can also be interlinked. The definition is open to add the effect of acquired components of which have a considerable effect on the target audience and vice versa.

1.2.2 Levels of Opinion Mining

Liu [19] has considered the research area that finds relevant information from analysis of human emotions, interest in different services and products, their opinion about product and services, attitudes, etc., as opinion mining. The three levels of opinion mining has given below:

Document Level: At this level, distinction of whole opinion document is done on the basis of two categories, positive category and negative category [8, 10]. This level has a pre-determined belief of relying on one entity, and single opinion holder constitutes the complete opinion. Subjective and objective categorization plays a significant role at document level. However, a relevancy issue occurs when the entire sentence might not contain a relevant point. We can use supervised as well as unsupervised methods of classification at document level. There are two approaches to do classification [20].

Supervised machine learning approach: In this type of classification, finite set of classes with the training dataset are available.

Unsupervised machine learning approach: Here, the polarity is calculated for each of the opinion words in the document, and accordingly, the document is said to have a positive score if the total polarity of these words comes with a positive count otherwise has a negative score [21].

There are various pros and cons of the document-level classification. The score of each individual sentence can be calculated, and limitation lies in the ignorance of feeling about the different attributes [22].

Sentence Level: Digging deeper into the documents than previous level, the complete sentence has been categorized into positive category, negative category, or neutral category at this level. Tough, neutral category resides empty every time. It is also related to subjectivity classification [9]. Every sentence is considered as a separate unit, and polarity is calculated for each sentence.

Feature Level: It finds a more detailed organization at aspect level. A quality based and advanced analysis distinguishes the aspects in the Web document. Feature level can be used interchangeably as a name of aspect level. Feature sentiment analysis is defined as analyzing every feature specified in the document or sentence [23]. Instead of considering language constructs like paragraphs, documents, clauses or phrases and sentences, it analyzes the features or views deeper into the document. Opinions will be categorized on the basis of the positive or negative object or target.

According to the proposed definition, 'opinion mining takes association of reviews as a key component and networks with the review sites and social media for marketing and selling their products and services.' The relationship with the components is maintained, which is emphasized by the common platform by substitution of resources, information, or artifacts. It can be concluded that this definition consider almost every aspect mentioned in the preceding definitions of opinion mining. However, opinion mining can be taken as set of business rules w.r.t. the e-commerce, somewhat true. The main goal of the business is to make profit by reviewing the sites, feedbacks, and people's view. The definition is correct if it is for individual organizational profit. However, our study is not based on this purpose. The goal of opinion mining should not be limited to make profit only. It is neither right nor ethical practice for the benefit of the society. So, the mechanism of opinion mining should achieve the task, and it is obligatory to achieve. It might intend to contribute toward a purpose of the on the whole mining system. In this sagacity, the analogy of IR would be more apposite than the commerce.

1.2.3 Components of Opinion Mining

Montoyo et al. and Cambria et al. [24, 25] have concluded their research that the main task of opinion mining should identify the opinion from different documents and elaborate them to the user automatically. The feature extraction of objects is the basis for the task of extracting public opinion.

Components
There are three categories of the components under the opinion mining that are distinguished below

- Object (Entity/Target)
- Opinion holder
- View.

Let us start with an example:
I bought a Samsung LED TV one week ago. I loved its features like the picture quality. It has smart television features. It has a big screen. The Wi-Fi slot enables me to watch my selection of movies from the Internet. However, my younger sister who is a fan of 3D cartoons was annoyed due to the absence of a 3D feature in the T.V. My mother wanted to return it back as it was too expensive.

This review example contains several opinions. Considering the view, positively reviewed sentences are 2, 3, 4, and 5, and negatively reviewed sentences are 6 and 7. The targets in this view are (1) television in sentence 1, (2) Wi-Fi slot in sentence 5, (3) sentence 2 contains picture quality, and (4) price in the sentence 7. The author is a holder in sentences 1,2,3,4, and 5. But in the last two sentences, i.e., 6 and 7, the holder is her sister and his mother. To make sound decisions, one has to pay the necessary heed to other's opinions. This is true for both individuals and organizations. The opinion mining application helps in taking decisions and making policies and extracting logic out of hundreds of intervention [26]. Opinions are gaining tremendous importance; therefore, the widespread use of opinion mining is the primary aid for various question answering systems, search engines, and other practical applications. The machine human interaction has also been enhanced by it. Opinion mining can be useful in several domains like in search engines, question answering systems, recommendation systems, etc. It also helps in developing better human computer communications [4, 13].

Tools
There are various tools available for opinion mining. Some of them are as follows:

Review Seer helps with the automation of the work through aggregation sites. The extracted feature terms are assigned a score using NB classifier by collecting set of positive and negative opinions. The output is in the form of review sentences with the score of each feature [27].

Web Fountain makes use of definite base noun phrase (dBNP) by extracting the product features following heuristic approach. Two resources which are used in this tool are sentiment lexicon and sentiment pattern database. Opinion sentences are parsed, and sentiments are provided for each feature with this tool [28].

Red Opal helps the user to determine the entities, and its polarity of the sentence is found on the basis of aspects. The orientation of the reviews is found on the basis of attributes from the opinions. The output is shown on a Web-based interface [29].

Opinion Observer is a sentiment analysis tool which displays output graphically by presenting comparison of the analysis made from the opinions posted online. It uses a WordNet exploring method to assign prior polarity.
    The following tools are used to find polarity from text.

Waikato Environment for Knowledge Analysis (Weka) is a Java suite of machine learning, used to compare the various machine learning algorithms on the basis of accuracy, precision, and recall for data mining applications [30].

Rapidminer helps in giving solutions and services in the area of analytics, data mining, and text mining applications.

Recall-Oriented Understudy for Gisting Evaluation (ROUGE) is the data summarization tool stated by Lin [31]. It is used to effectively calculate the worth of summary generated between humans and systems. The overlapping units are counted among both summaries such as the N-gram, word sequences, and word pairs.

NLP tools: General Architecture for Text Engineering (GATE) is NLP and language engineering tool. Natural Language Toolkit (NLTK), Ling Pipe, Open NLP, and Stanford Parser and POS Tagger, etc. [20], are part of NLP based tools.

1.2.4 Applications in Opinion Mining

The major application areas are buying or selling products or services, quality control areas, policy or decision making, business intelligence by conducting market research, and so on. Applying sentiment analysis on the feedbacks received by people automatically provides valuable information for further analysis on market reports [32]. The major areas which have the highest impact of opinion mining are:

- Buying and quality enhancements in services and products by inserting advertisements in user-generated content.
- Tracking political topics for decision making and making policy.
- Optimized opinion search results using content analysis automatically.

In general, an image of certainty can be used interchangeably for opinion mining as it pays attention through listening instead of asking for recognizing the difficulties. It requires great number of approaches of sentiment classification into positive, negative, or other categories.

1.3 Steps in Opinion Mining

Opinions can be considered to be an assessment of an object, feeling, sentiments, product or entity, organization or event, and a review [33]. The term object can be defined as an entity on which an opinion is given. It has a hierarchical relationship, which is represented as a set of attributes such as Obj: (C, SubC), where C represents a hierarchy of components and SubC denote subcomponents of the object Obj. The term feature or aspect represents the components that define the objects. It has associations with the components of objects or its subcomponents. The steps in opinion mining are given in Fig. 1.3.

For example, a document D consists of sentences taken as opinions o_1, o_2, o_3, and so on.

These individual opinions consist of aspects which specifically name the particular words that clearly state the view of an opinion holder which is expressing their opinion on $D = \{o_1, o_2, o_3 \ldots\}$. $o_1 = \{a_1, a_2, a_3 \ldots\}$, where a_1, a_2, or a_3 represent features.

The steps followed in the above figure are explained as follows.

Fig. 1.3 Steps in opinion mining

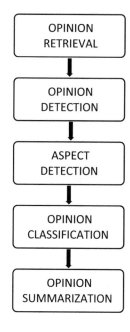

1.3.1 Opinion Retrieval

The Web crawler is used for gathering opinions from different sources such as blogs, forums, e-news, review channels, tweets, and social networking sites with the help of the Web crawler [34].

1.3.2 Opinion Detection

Four important parameters are to be considered for detecting opinions (trust, polarity, opinion score, and quality) stated as: Is the document relevant? Is the document of good quality? Is the document carrying the opinion or not? Does the opinion fall under spam detection?

Currently, the following methods are used in opinion detection are:

- Lexicon-based method:—which uses frequency of certain frequent terms from text to rank documents.
- Shallow linguistic approach:—based on frequency of pronouns and adjectives.
- Machine learning:—text level compressibility technique, keyword stuffing, supervised learning methods such as support vector machines (SVM), maximum entropy (ME), naïve Bayes (NV), and logistic regression (LR).

1.3.3 Aspect Detection

The task of 'opinion mining' is to extract people's opinion on features of an entity which can be done on the three levels:

- Document level: Classify the whole document into different categories opinions like positive or negative.
- Sentence level: Analyze the document at sentence level where sentences are classified into positive, negative, or neutral. Knowing some aspect in the sentence is important.
- Feature level: Deal with identifying features in sentences for a document and analyzing features into positive, negative, or neutral.

The present methods are supervised learning methods such as SVM, ME, NB, LR which uses unigram, bigram, word data, meta-data, affix similarity, and word emotion. Unsupervised learning methods use bootstrapping, rule-based approach, or probabilistic-based approaches. Semi-supervised learning methods are also available for finding features from opinions.

1.3.4 Opinion Classification

The other terms associated with opinion classification is called as 'opinion orientation' or 'semantic orientation.' Each opinionated word is analyzed, and polarity is determined as positive, negative, or neutral. The present methods are:

- Endogenous method—These methods estimate and predict the polarity of words in a document using machine learning techniques.
- Exogenous methods—These methods use lexical approach which analyzes individual words and/or phrases of sentiment, and refer emotional dictionaries: emotional lexical items from the dictionary are searched in the text, their sentiment weights are calculated, and some aggregated weight function is applied [35].

1.3.5 Opinion Summarization

There are different means for interpretation of summary of opinions. The important sentences are combined together in order to give a meaningful summary. Two main approaches are:

- Extractive summarization is the technique of concatenating extracts taken from a corpus into a summary.
- Abstractive summarization involves paraphrasing the corpus using novel sentences.

Other methods are:
Feature-based method, term frequency-based method, and color representation method [36]. An example is given in Fig. 1.4.

Fig. 1.4 An example of summary of opinions based on features

iPhone 7:

PHONE:

Positive: 195	<opinions>
Negative: 81	<opinions>

Feature: battery life

Positive: 63	<opinions>
Negative: 75	<opinions>

Feature: picture quality

Positive: 132	<opinions>
Negative: 6	<opinions>

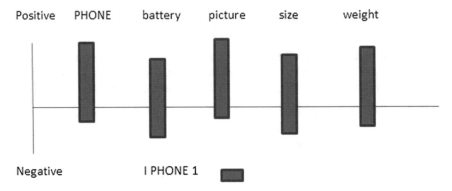

Fig. 1.5 Summary of opinions (iPhone) based on aspects

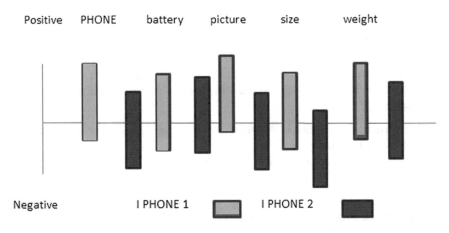

Fig. 1.6 Opinion comparison of two iPhones

Two aspects, 'battery life' and 'picture quality,' are also shown. In each of the aspects, the positive reviews and negative reviews are taken. It has 132 positive reviews, and 6 negative reviews for the aspect 'picture quality.' We have visualized the above example in the bar chart also. Figure 1.5 shows two axes, the above x-axis give the positive reviews, and below x-axis gives negative reviews. The comparison of the features of two different iPhones is illustrated in Fig. 1.6.

1.4 Information Retrieval: Challenges in Mining Opinions

A complete vision of the overall work that has been accomplished under the different areas under opinion mining is explained in detail in the above sections. The novel algorithms and techniques that have been proposed with their variations and advantages are highlighted. The various limitations that are analyzed while going into the

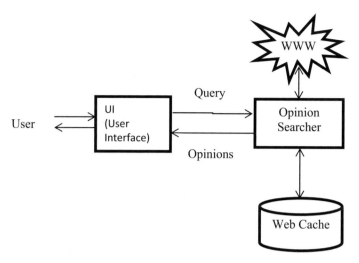

Fig. 1.7 Model of IR opinion system

depth of literature are discussed in this chapter which can be further considered as the future scope for researchers. The keen motivation to develop the application of social network analysis to Web IR systems has been emphasized. This deep finding enlightens to provide insight into understanding of the opinions and interpret relations between documents and individuals to improve the relevant retrieved documents as a response to the user query. Merging information from review sites into the information extraction process helps in developing a Web IR system [4]. This book relies on three components corresponding to each entity in the information model. Figure 1.7 shows how opinions and other social network information are used for designing an IR system.

WWW is considered as a treasure of massive information that gives knowledgeable information for day-to-day applications. IR systems provide us a way to traverse through this enormous amount of information available at WWW. The gaining attractiveness of Facebook, Twitter, and other sites involved in social media [5] on WWW are majorly involved in providing good exposure to Web intelligence and IR systems [37]. Since the last decade, persons engaged in social media have been raised from 127 million to 283 million from the year 2013–2017. This growth has created the dependency of users on IR systems and search engines. So, an emergent explicit need is to be fulfilled for the people in developing an IR system that will help them to analyze the opinions as positive or negative and view the concise summary considering aspects relevant to the particular domain.

1.4.1   Introduction to IR

Information retrieval is 'finding material (usually documents) of an unstructured nature (usually text) that satisfies an information need from within large collections (usually stored on computers).' IR is extracting important patterns, features, knowledge from data. In other words, it can be defined as 'facts provided or learned about something or someone.' IR system is a network of algorithms, which facilitate the search of relevant data/documents as per user requirement.

1.4.2   Present Opinion Mining Systems

Hu and Liu [3, 38] proposed an opinion mining system called feature-based summarization using computational linguistics. Opinion words were identified using simple NLP linguistic parser for finding the POS for each word. The approach adopted by [39] to find the adjectives as opinion words by identifying nouns was used. The polarity of sentences was determined using WordNet[1] dictionary and bootstrapping technique. The traditional text summarization technique was used by selecting a subset of the original sentences from the reviews to capture their main points. Another similar system was proposed by Liu and Hu which focused on specific features of the product that customers have opinions on. The text summarization technique proposed earlier was different as it worked on specific features and summarizes all the customer reviews of a product [3]. A system called 'OPINE' was developed by Popescu et al. [40] which used unsupervised learning approach for identifying features and opinions and ranked the opinions by calculating their strength of the reviews. The summarization of Twitter messages based on Lucene formula was purely application and formula based. The system developed using the older method of scoring the documents using WordNet score and lacked features like aspect detection, etc. Wani and Patil [41]. Really simple syndication system [42] was proposed, and it was based on blogs which relies on analysis done by humans, and scores are calculated using an online dictionary. The targeted opinion words based on the features with their adjectives require domain-specific knowledge for producing appropriate results. Sumview system was proposed by Wang et al. [43] which used clustering technique for the grouping of reviews based on feature-based weighted nonnegative matrix factorization (NNMF) algorithm. The summary was based on choosing the most appropriate opinion sentences according to the aspect selected respective to the particular domain.

Earlier developed opinion mining systems have certain disadvantages:

- Earlier methods of extracting reviews discuss parsing of the Hypertext Markup Language (HTML) code of the Web page. As these Web pages frequently undergo updations and have fresh opinions, techniques must be developed which would

---

[1]WordNet is a freely and publically available large lexical dataset of English.

crawl the reviews and calculate the revisit frequency of each Uniform Resource Locator (URL) periodically to extract relevant and fresh opinions.

- Nouns and its associated closest adjectives method were the most approachable criteria of identifying aspects. Ontology-based aspect detection involves populating the database with the semantic information which has been widely adopted previously for finding aspects. Many different approaches used in aspect identification produced higher precision for a specific domain but low recall, and on the other-hand, some approaches produced higher recall with low precision. Therefore, a combination of different approaches can yield better results. Novel hybrid algorithms need to be proposed for identifying features.
- For classifying the sentiments in 'AskUs: Opinion search engine' [44], the max score algorithm defining the parameter called as 'weight' needs to be tuned which may give inaccurate results for varying data sets. Leaving the traditional methods of using SentiWordNet[2] that assigns polarity scores to each word, novel techniques must be adopted that would learn features on its own and adjust weights automatically though training.
- Summary of sentences did not include sentiment fusion. Repetitive sentences should be avoided, and different opinions must be combined in order to present more concise form of result.

1.4.3   Challenges: Factors that Make Opinion Mining Difficult

Opinion mining lies at the crossroads of computational linguistics and IE. Therefore, proposed work combines theories of supervised learning techniques with NLP tasks in order to develop the complete IR system. The various researchers [6, 19, 20, 23, 45] have discussed the efferent concepts of opinion mining including its meaning, tasks, techniques, challenges, sentiment analysis, and other aspects in summarization of opinions. Penalver-Martinez et al. [46] converse about role of opinion mining in semantic Web technologies. Song et al. [47] have proposed and presented a novel technique for extracting opinion by assessing product reviews. The authors Dongre et al. [48] anticipated a novel model and a generic tool for aspect-based summarization and classification of reviews such that it does not entail any pre-requisite information or seed words about services and products. A tool has been presented by the authors [49] to accumulate and visualize data from popular social media Web site Facebook. Another facet of opinion summarization based on features of acquired reviews has discussed in detail by Bafna and Toshniwal [50]. Sentiment analysis is achieved by using supervised and unsupervised learning techniques, and evaluation is discussed by involving performance metrics on different datasets. The authors

---

[2]SentiWordNet is a lexical resource for opinion mining. SentiWordNet assigns to each synset of WordNet three sentiment scores: positivity, negativity, objectivity (http://SentiWordNet.isti.cnr.it/).

[51] have given an aspect-based opinion mining technique by hybrid theories, and results were evaluated by effective measures using information retrieval (IR) evaluation techniques. The authors [10] have applied machine learning algorithms and presented out the effective solution to sentient analysis problem. The researchers, Hemalatha et al. [52], developed a tool for sentiment analysis of tweets from a popular social media Web site. Some other researchers have identified sentiments at all levels and analyzed its subcomponents [53]. The detailed analysis is presented on tweets using three metrics names as precision, recall and F-measure.

Early years of research in opinion mining focused on precise content. Therefore, there were modest emphasis on a consistent structure—no integrative framework has been proposed for mining opinions from documents. Furthermore, a less attention has given on standardizing the structure of process of mining opinions. The constant efforts made by many institutes for researching in mining the opinions can be enlisted as one of the major resources of creating a massive quantity of online reviews and feedbacks of different products.

Existing opinion mining systems typically depend on the way the reviews are collected, the type of aspects which are extracted, and the pattern followed to find the polarity of the sentiments [54, 55]. People's opinions, which are extracted from the writings on the Web, largely depend on the above-mentioned factors, and the summary of opinions formed should be structured and concisely in the aggregated form covering the important aspects of the products.

1.4.4   Our Charge and Approach

The opinions are extracted from seed URL's, and pre-processing operations are applied on the opinions. Grammatical relations are explored which are used to find the relationship between target entities and features on the opinions. Examples of opinion words are poor, beautiful, great, etc., and target entities are battery, picture, speed, etc. Then, the semantic orientations of opinions are determined based on the aspects identified. Sentiment classification at the aspect level is the next prime task, which classifies each sentence based on the aspects into positive and negative category. The algorithm of deep learning, i.e., CNN is applied on the opinions for categorizing opinions. Finally, ranking and summarization of opinions are achieved using the PCA technique, and abstractive summary is generated using graphs. Various datasets are available which will be considered in this book to compare the results with the existing work.

The book is supposed to build an IR opinion mining system with the following Objectives:

- The existing Web search engine is used to develop an opinion retriever for extracting opinions. The novel algorithms are proposed in order to detect Web pages which frequently undergo updation by calculating the time for its revisit in order to extract relevant and fresh opinions. The extraction and ranking of objects of interest are done after applying pre-processing tasks.

- The aspects are identified by proposing rules based on grammar dependency. After detecting aspects using NLP tools and dependency relations (extracting opinion word feature pair like amazing battery, pretty ugly voice quality, etc.), sentiment analysis is performed.
- The sentences are categorized into one of the two categories positive and negative based on the learning algorithms of deep learning. Classification of opinions is done by selecting appropriate parameters and configurations using CNN.
- The abstractive summarization of sentences is achieved by exploiting the graph structure of a sentence using NLP approaches. The extractive summary is formed using PCA. Adequately, comparison with both of the techniques is performed, and results of the same are analyzed and compared using recall-oriented understudy for gisting evaluation (ROUGE) tool.

This book addresses combination of NLP and deep learning theories to solve and optimize the problem of classification of sentiments and to develop an efficient opinion system. Comparative experiments on different data sets which contain real discussions and reviews are conducted, and the accuracy is effectively measured using performance metrics and data mining tools.

1.5   Summary

This chapter discusses some background knowledge relevant for the book. The work focuses on developing an IR system for constructing a sentiment model for fine-grained opinion mining from unstructured content in online shopping sites. Therefore, analysis requires rendering an insight on the terms relevant to this work for deeper understanding of the opinion mining system. The retrieval of opinions requires traversing social interaction sites and various online shopping sites. Richer modeling and mining of opinions lie in finding the important features from opinions, and it involves natural language processing of text. Classifying the opinions into positive or negative classes can be done using machine learning algorithms or using deep learning algorithms. Finally, the overall summary of opinions is presented using different summarization techniques. The basic model of opinion mining and its development process are hereby discussed.

References

1. Shu, H. (2010). Opinion mining for song lyrics. Thesis.
2. Sundheim, B. M. (1996). The message understanding conferences. In *Proceedings of a workshop on held at Vienna* (pp. 35–37). Virginia: Association for Computational Linguistics.

3. Hu, M., & Liu, B. (2004) Mining opinion features in customer reviews. In *AAAI 2004* Jul 25 (Vol. 4, No. 4, pp. 755–760).
4. Kirchhoff, L. (2010). Applying social network analysis to information retrieval on the World Wide Web: A case study of academic publication space.na.
5. Jarke, M., & Klamma, R. (2007). Social software und Reflektive Informationssysteme. In *Architekturen und Prozesse* (pp. 51–62). Berlin: Springer.
6. Lo, Y. W., & Potdar, V. (2009). A review of opinion mining and sentiment classification framework in social networks. In *2009 3rd IEEE international conference on digital ecosystems and technologies* (pp. 396–401). IEEE.
7. Dave, K., Lawrence, S., & Pennock, D. M. (2003). Mining the peanut gallery: Opinion extraction and semantic classification of product reviews. In *Proceedings of the 12th international conference on World Wide Web* (pp. 519–528). ACM.
8. Turney, P. D. (2002). Thumbs up or thumbs down?: Semantic orientation applied to unsupervised classification of reviews. In *Proceedings of the 40th annual meeting on association for computational linguistics* (pp. 417–424). Association for Computational Linguistics.
9. Wiebe, J. M., Bruce, R. F., & O'Hara, T. P. (1999). Development and use of a gold-standard data set for subjectivity classifications. In *Proceedings of the 37th annual meeting of the association for computational linguistics on computational linguistics* (pp. 246–253), Association for Computational Linguistics.
10. Pang, B., Lee, L., & Vaithyanathan, S. (2002, July). Thumbs up?: sentiment classification using machine learning techniques. In *Proceedings of the ACL-02 conference on empirical methods in natural language processing-Volume 10* (pp. 79–86). Association for Computational Linguistics.
11. Abulaish, M., Doja, M. N., & Ahmad, T. (2009). Feature and opinion mining for customer review summarization. In *Pattern recognition and machine intelligence* (pp. 219–224). Berlin: Springer.
12. Seerat, B., Azam, F. (2012). Opinion mining: Issues and challenges (A survey). *International Journal of Computer Applications* (0975 – 8887), *49*(9).
13. Khan, K., Baharudin, B., Khan, A., & Ullah, A. (2014). Mining opinion components from unstructured reviews: A review. *Journal of King Saud University-Computer and Information Sciences, 26*(3), 258–275.
14. Liu, B. (2008). Opinion mining. Encyclopedia of Database Systems.
15. Ganesan, K. A., & Kim, H. D. (2008). *Opinion mining-A short tutorial (Talk)*. University of Illinois at Urbana Champaign.
16. Sharma, N. R., & Chitre, V. D. (2014). Opinion mining, analysis and its challenges. *International Journal of Innovations & Advancement in Computer Science, 3*(1), 59–65.
17. Mobasher, B. (2007). Data mining for web personalization. In *The adaptive web* (pp. 90–135). Berlin: Springer.
18. Ma, Z. M. (2005). Engineering information modeling in databases: needs and constructions. *Industrial Management & Data Systems, 105*(7), 900–918.
19. Liu, B. (2012). Sentiment analysis and opinion mining. *Synthesis lectures on human language technologies, 5*(1), 1–67.
20. Kaur, J., & Saini, J. R. (2015). A study of text classification natural language processing algorithms for indian languages. *The VNSGU Journal of Science Technology, 4*(1), 162–167.
21. Kolkur, S., Dantal, G., & Mahe, R. (2015). Study of different levels for sentiment analysis.
22. Devi, G. D., & Rasheed, A. A. (2015). A survey on sentiment analysis and opinion mining. *International Journal for Research in Emerging Science and Technology, 2*(8), 26–31.
23. Kaur, A., & Duhan, N. (2015). A survey on sentiment analysis and opinion mining. *International Journal of Innovations & Advancement in Computer Science, 4*, 107–116.
24. Cambria, E., Schuller, B., Xia, Y., & Havasi, C. (2013). New avenues in opinion mining and sentiment analysis. *IEEE Intelligent Systems, 28*(2), 15–21.
25. Montoyo, A., MartíNez-Barco, P., & Balahur, A. (2012). Subjectivity and sentiment analysis: An overview of the current state of the area and envisaged developments.

26. Osimo, D., & Mureddu, F. (2012). *Research challenge on opinion mining and sentiment analysis* (p. 508). Bâtiment: Universite de Paris-Sud, Laboratoire LIMSI-CNRS.
27. Maynard, D., Bontcheva, K., & Rout, D. (2012). Challenges in developing opinion mining tools for social media. In *Proceedings of the@ NLP* (pp. 15–22).
28. Gruhl, D., Chavet, L., Gibson, D., Meyer, J., Pattanayak, P., Tomkins, A., & Zien, J. (2004). How to build a WebFountain: An architecture for very large-scale text analytics. *IBM Systems Journal, 43*(1), 64–77.
29. Scaffidi, C., Bierhoff, K., Chang, E., Felker, M., Ng, H., & Jin, C. (2007). Red opal: Product-feature scoring from reviews. In *Proceedings of the 8th ACM conference on electronic commerce* (pp. 182–191). ACM.
30. Holmes, G., Donkin, A., & Witten, I. H. (1994). Weka: A machine learning workbench. In *Proceedings of the 1994 second Australian and New Zealand intelligent information systems conference* (pp. 357–361). IEEE.
31. Lin, C. Y. (2004). Rouge: A package for automatic evaluation of summaries. In *Text summarization branches out: Proceedings of the ACL-04 workshop* (Vol. 8).
32. Smeureanu, I., & Bucur, C. (2012). Applying supervised opinion mining techniques on online user reviews. *Informaticaeconomica, 16*(2), 81.
33. Bhatia, S., Sharma, M., Bhatia, K. K. (2015). Strategies for mining opinions: A survey. In *2015 2nd international conference computing for sustainable global development (INDIACom)* (pp. 262–266).
34. Bhatia, S., Sharma, M., Bhatia, K. K., & Das, P. (2018). Opinion target extraction with sentiment analysis. *International Journal of Computing, 17*(3), 136–142. (Elsevier, SCOPUS).
35. Blinov, P., Klekovkina, M., Kotelnikov, E., & Pestov, O. (2013). Research of lexical approach and machine learning methods for sentiment analysis. *Computational Linguistics and Intellectual Technologies, 2*(12), 48–58.
36. Fukumoto, F., & Suzuki, Y. (2000). Event tracking based on domain dependency. In *Proceedings of the 23rd annual international ACM SIGIR conference on research and development in information retrieval* (pp. 57–64). ACM.
37. Mika, P. (2005). Ontologies are us: A unified model of social networks and semantics. In *International semantic web conference* (pp. 522–536). Berlin: Springer.
38. Hu, M., & Liu, B.(2004). Mining and summarizing customer reviews. In *Proceedings of the tenth ACM SIGKDD international conference on knowledge discovery and data mining*. ACM.
39. Miller, G. A. (1995). WordNet: a lexical database for English. *Communications of the ACM, 38*(11), 39 36b.
40. Popescu, A. -M., Nguyen, B., & Etzioni, O. (2005). OPINE: Extracting product features and opinions from reviews. In *Proceedings of HLT/EMNLP on interactive demonstrations*. Association for Computational Linguistics.
41. Wani, S. S., & Patil, Y. N. (2015, June). Analysis of data retrieval and opinion mining system. In *2015 IEEE international advance computing conference (IACC)* (pp. 681–685). IEEE.
42. Al-Hamami, A. H., & Shahrour, S. H. (2015, December). Development of an opinion blog mining system. In *2015 4th international conference on advanced computer science applications and technologies (ACSAT)* (pp. 74–79). IEEE.
43. Wang, D., Zhu, S., & Li, T. (2013). Sumview: A Web-based engine for summarizing product reviews and customer opinions. *Expert Systems with Applications, 40*(1), 27–33.
44. Pisal, S., Singh, J., & Eirinaki, M. (2011, December). AskUs: An opinion search engine. In *2011 IEEE 11th international conference on data mining workshops (ICDMW)* (pp. 1243–1246). IEEE; Smolander, J. (2016). Deep learning classification methods for complex disorders.
45. Vinodhini, G., Chandrasekaran, R. M. (2012). Sentiment analysis and opinion mining: a survey. *International Journal, 2*(6).
46. Penalver-Martinez, I., Garcia-Sanchez, F., Valencia-Garcia, R., Rodríguez-García, M. Á., Moreno, V., Fraga, A., et al. (2014). Feature-based opinion mining through ontologies. *Expert Systems with Applications, 41*(13), 5995–6008.
47. Song, Q., Ni, J., & Wang, G. (2013). A fast clustering-based feature subset selection algorithm for high-dimensional data. *IEEE Transactions on Knowledge and Data Engineering, 25*(1), 1–4.

48. Dongre, A. G., Dharurkar, S., Nagarkar, S., Shukla, R., Pandita, V. (2016). A survey on aspect based opinion mining from product reviews. *International Journal of Innovative Research in Science, Engineering and Technology, 5*(2), 2319–8753. ISSN(Online).

49. Mfenyana, S. I., Moroosi, N., Thinyane, M., & Scott, S. M. (2013). Development of a Facebook crawler for opinion trend monitoring and analysis purposes: case study of government service delivery in Dwesa. *International Journal of Computer Applications, 1, 79*(17).

50. Bafna, K., & Toshniwal, D. (2013). Feature based summarization of customers' reviews of online products. *Procedia Computer Science, 22*, 142–151.

51. Varghese, R., Jayasree, M. (2013). Aspect based sentiment analysis using support vector machine classifier. In *International conference on advances in computing, communications and informatics (ICACCI)*, IEEE.

52. Hemalatha, I., Varma, D. G., & Govardhan, A. (2013). Sentiment analysis tool using machine learning algorithms. *International Journal of Emerging Trends & Technology in Computer Science (IJETTCS), 2*(2), 105–109.

53. McDonald, R., Hannan, K., Neylon, T., Wells, M., Reynar, J. (2007) Structured models for fine-to-coarse sentiment analysis. In *Annual meeting-association for computational linguistics 2007* Jun 23 (Vol. 45, No. 1, p. 432).

54. Dey, N., Babo, R., Ashour, A. S., Bhatnagar, V., & Bouhlel, M. S. (Eds.). (2018). *Social networks science: Design, implementation, security, and challenges: From social networks analysis to social networks intelligence*. Berlin: Springer.

55. Singh, A., Dey, N., Ashour, A. S., & Santhi, V. (Eds.). (2017). *Web semantics for textual and visual information retrieval*. IGI Global.

Chapter 2
Opinion Score Mining System

Researchers found that the opinions of people are scattered on various review sites and social networking sites, so instead of traversing the different web sites for the reviews with their polarity education of sentences, a complete dynamic IR system should be developed that gives ranked opinions on the basis of features in summarized form to the users to make much quicker and better decisions in real time. In this chapter, an organized and unified structure that accounts for mining opinions in a more general and easier way is discussed. The proposed work has been drawn on previous researches already achieved in the separate areas done under opinion mining, i.e., extracting opinions, classifying opinions, and summarizing opinions. The proposed Opinion Score Mining System (OSMS) discusses three major steps that constitute the opinion mining system. The work starts with crawling of opinions and cleaning them by using different pre-processing tasks. The opinion retriever discussed in Chap. 3 extracts the most relevant and recent reviews, periodically update them by calculating the precedence of the seed URLs. Features are identified and the thematic opinion word pair is formed by carefully selecting grammatical relations. Next, the opinions are segregated into the classes, namely positive and negative by using the algorithm of deep learning. Selectively, choosing the hyperparameters and configuring the basic CNN model, better results are achieved in with opinion classification when compared with previous models. The summary is generated using both abstractive and extractive techniques. The abstractive summary is also formed by using novel algorithms proposed for fusing sentiments together, thoroughly checking grammar of the sentences, and removing repetitive sentences. Extractive summary is generated by implementing PCA on the matrix created using SentiWordNet. The analysis of both the techniques is conducted using ROUGE tool and results are evaluated. As the initial analysis in opinion mining did not find much evidence in stating that there exists a unique system for summarizing opinions; therefore, theories are suggested in proposing hybrid techniques and novel methodologies for presenting consolidated aspect-based summary of opinions. Various experiments are conducted

© The Author(s), under exclusive license to Springer Nature Singapore Pte Ltd. 2020
S. Bhatia et al., *Opinion Mining in Information Retrieval*,
SpringerBriefs in Computational Intelligence,
https://doi.org/10.1007/978-981-15-5043-0_2

using different techniques on real datasets and standard datasets which justifies that
the notion of developing such a system is necessary and useful to build.

2.1   Framework Design

WWW is a treasure of reviews obtained by different persons. In this chapter, the
emphasis is on describing the opinion system as a framework that gives ranked opinions on various domains as demanded the users. The basic structure of opinion mining
and the steps are explained in detail in this chapter. The overall flow of mining opinions is described and results are evaluated on the secondary datasets in terms of IR
evaluation measures. This system defines the development of the complete architecture to mine opinions, providing detailed description of each component, including
algorithms, definitions, and behavior. Without explaining each step in detail, it would
be difficult to understand how the components would work and behave in mining
opinions. On the other hand, with a set of definitions and behavior descriptions, more
informed decisions can easily be made by people whether to buy a certain product or
not? When the user wishes to find opinions related to electronics domain, this will
fulfill his/her need in the form of relevant ranked opinions in the final summarized
form. The functional modules are explained in detail below:

The framework design is shown in Fig. 2.1.

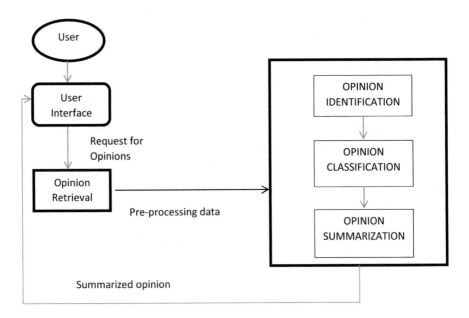

Fig. 2.1 Framework design

2.1.1 Opinion Retrieval

The task of opinion retrieval is to extract reviews from various review sites and social networking sites. The seed URLs are obtained from the several search engines to retrieve the opinions. In this chapter, 'Google' search engine is used due to its speed retrieval process and its precision-based text results [1]. Several novel algorithms are proposed to get the most recent opinions. Separate repository will be created to store the opinions fetched from the existing retriever with its appropriate date and time. Dynamic priority for every seed URL will be computed and the time of its revisit will be calculated to maintain freshness of the opinions. The reviews collected will be pre-processed for eliminating noise by applying data pre-processing tasks like Tokenization, Lemmatization, Stop word removal, etc. The major research work focuses on analyzing aspects from opinions, classifying opinions, and summarizing them as highlighted in Fig. 3.1. Primarily, three tasks will be considered as follows:

Identification of aspects, classifying the opinions into positive and negative category, and finally, summarizing them to present the overall summarized opinions based on the features.

2.1.2 Opinion Identification

After pre-processing is done, relevant content needs to be extracted and features need to be identified first for classifying opinions. Since, the chapter focuses on considering aspect-based opinion summarization, extracting features is the prime objective. Previous researches have been done on extracting features based on nouns and its phrases and also on relations between them. The novel method based on grammar dependency rules is proposed for aspect extraction. For every review, thematic word and related opinion word are extracted. The thematic word is the target entity (feature) that is related to the review and the latter deals with identifying the associated adjective with the aspect.

2.1.3 Opinion Classification

The opinions will be classified according to the target entity into either positive or negative category. The deep learning algorithm, i.e., CNN is used for classifying the sentiments. The accuracy of different classifiers is compared with CNN by modifying the hyperparameters.

2.1.4   Opinion Summarization

The final summarization is formed by presenting the concise view including all relevant points related to that product. For summarization, as mentioned before, two techniques are proposed. The first technique uses graphs-based methodology with the application of NLP. The second method is accomplished by preparing a matrix, consulting opinion word repository with SentiWordNet score, implementation of PCA by applying SVD, and finally, sorting the sentences above threshold which will form the basis for extractive summary. The comparison is done and detailed analysis is carried out with both the techniques by calculating ROUGE scores.

2.2   Opinion Score Mining System (OSMS)

This system responds to the user's query with the set of opinions returned as the search results together with the overall summary of the opinions. The proposed OSMS includes several steps. A critical look at the available literature and keeping the observations made thereof in mind, it is felt that a survey needs to be conducted to find the requirements of the users from a general search engine (e.g., Google, Yahoo, AltaVista, etc.) and to study whether general purpose search engines are able to fulfill the requirements of users or not? A survey was conducted in different universities, namely YMCA University of Science and Technology,[1] Faridabad, Echelon Institute of Technology,[2] Faridabad and KR Mangalam University[3] among 60 people in total comprising of students, teachers, technical, and non-teaching staff. The survey consisted of 15 different products on electronics domains and opinions. For each product, a number of positive and negative opinions were given. After conducting the survey on a few products, all the responses were collected and they are given in Table 2.1 shows the details where Product-id is the product enquired in the survey and the reviews provided is given by positive and negative opinions.

In analyzing the responses, the following observations have been made:

- It has been observed that most of the users felt that, for finding the summarized opinion based on its aspects, the proposed OSMS can offer much better results than the traditional search engine.
- It is further observed that different Web pages returned by the search engines lead to searching of the reviews which is a cumbersome task and getting appropriate summarization related to the aspects on which a person may take the right decision is far from what is expected.

It has also been observed that a large number of reviews have been written by users on review sites and social media, and they play a crucial role in providing answers

---

[1] http://ymcaust.ac.in/.

[2] http://www.eitfaridabad.com/.

[3] https://www.krmangalam.edu.in/.

Table 2.1 Survey responses

Product ID	Positive	Negative	Neutral
1	57	3	–
2	20	19	21
3	45	15	–
4	40	13	7
5	35	10	15
6	40	10	10
7	38	22	–
8	10	43	7
9	8	42	9
10	40	20	–
11	38	5	17
12	38	10	12
13	47	13	–
14	45	15	–
15	10	43	7

corresponding to user's query on that topic. A graph has been plotted for the responses received by the participants as shown in Fig. 2.2.

Keeping in mind the need of the information retrieval system using opinion mining, a framework for the OSMS is being proposed that can provide opinions to user's queries by providing opinions categorically and in summarized ranked form. The book emphases on three major functionalities: extracting relevant aspects from opinions, categorizing them into positive and negative categories, and then, summarizing

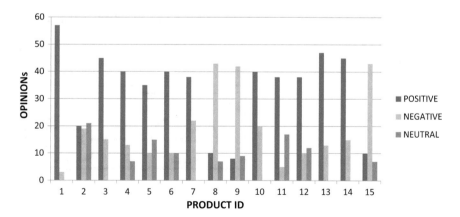

Fig. 2.2 Graph showing the responses of survey conducted

them. The user asks a question and gets opinion(s) in response. The detailed block diagram of the OSMS has been discussed in the next section.

The proposed design of a novel IR system for prospective opinion mining is shown in Fig. 2.3. It takes the input from the user in form of queries and returns the opinion(s) from its resources in the summarized form. The various algorithms and techniques used in the following steps of OSMS are discussed in the further chapters. The steps given below are explained briefly.

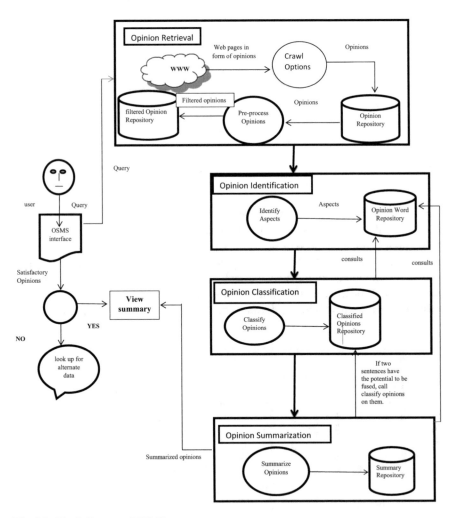

Fig. 2.3 Block diagram of OSMS

- Opinion crawling and pre-processing opinions
- Aspect identification and classification of opinions
- Aspect-based opinion summarization
- Look up for alternate data

A brief discussion on each of these functional components is given below:

2.2.1 Opinion Crawling and Pre-processing Opinions

The first part focuses on extracting opinions. Opinion retriever is of the major component of the proposed architecture that downloads the Web pages from the WWW. The functionality of opinion retriever is similar to the general crawler with a major difference that it downloads the opinions only, unlike the general crawler that downloads all types of Web pages. Since the opinions posted gets updated at a greater frequency than the general Web pages, the retriever needs to recrawl the Web pages more frequently to maintain the freshness of the repository. It is possible by applying more appropriate time interval between two successive crawls of the Web pages [2]. The process of crawling starts with an initial set of URLs possibly the top N among the recent review sites and social networking sites. The following steps given below will be repeated for seed URLs:

1. Extract the URLs from the list of seed URLs.
2. Check whether the URL Web pages contain opinions or not?
3. If yes, the Web page is downloaded from WWW and then stored in a buffer.
4. Fetch the relevant opinions by parsing the HTML code using basic string processing and indexes the opinions and store them further in the opinion repository with the link information.
5. The freshness of the opinions in the opinion repository is maintained by computing the revisit frequency of the Web pages.
6. Extract the links existing in the opinion pages and add in the list of seed URLs for further crawling.

It is worth noting that there are some features of opinion pages that are different from the features of the general Web pages, as listed below:

- There is an ordering among the Web pages containing opinions such that the opinions of the URLs which get updated more often will make the revisit rate of that URL to increase.
- The URLs of the Web pages usually contain a review, comment, and opinion summary as text keywords in the <span> tag or <div> tag, appears in the URL.
- The revisit frequency of the retriever by using the following formula:

$$\Delta t = D_{\text{new opinion}} - D_{\text{opinion}} + T_{\text{new opinion}} - T_{\text{opinion}} \qquad (2.1)$$

where, D refers to as Date in DD/MM/YY format and T refers to as Time in Hours: Min (24 h) format.

- The opinion retriever incrementally refreshes the existing collection of pages by visiting them frequently by calculating the time difference of the opinions of each URL.

Keeping these features in mind, architecture for the opinion retriever is designed and implemented. The detailed discussion with experimental analysis of the proposed architecture is given in Chap. 3. The opinion extracted undergo with the cleaning task in order to remove irrelevant content and make the opinions suitable for performing other tasks. The opinion repository is read and stored with the link information.

The following are the pre-processing tasks [3] applied to the opinions and are stored in the filtered opinion repository.

- Tokenization: It is referred to as process of breaking the sentences into small subparts known as tokens.
- Stop word elimination: The words which occur frequently in a sentence like a, an, the, etc., are removed as they are considered irrelevant and does not contribute in this chapter.
- Lemmatization: This is used to reach to the root word of the sentence by considering the grammar of the words. It includes the removal of derivational affixes.

For POS tagging the documents, Stanford NLP Parser is used [4, 5] which helps in splitting up into tokens and generates POS tagged Extensible Markup Language (XML) document as an output file. For example, for the sentence 'The phone has bad screen' obtains the following lemmatized words accompanied by their grammatical categories. The tokens with their tags are shown in Table 2.2.

2.2.2 Aspect Identification and Classification of Opinions

Aspect extraction involves extracting the features from the opinion. This involves participation of extracting the targeted opinion word pair and classifying opinions based on the identified opinion words [6, 7]. The aspects are identified using syntactic

Table 2.2 Tokenized opinions

Token	Tag
The	Determiner
Phone	Noun
Has	Verb
Bad	Adjective
Screen	Noun

Table 2.3 List of thematic-opinion word pair

Opinions containing aspects	Identified aspects
The laptop has an amazing battery life	Amazing battery life
The voice quality of the phone seems to be pretty ugly	Pretty ugly voice quality
My mother got angry with the price	Angry price
The lens is too much expensive to purchase	Too much expensive lens
The glass is too comfortable to wear	Too comfortable glass
I had a lovely time with you, enjoyed the movie	Enjoyed movie
This is useless glass	Useless glass

rules of natural language. The aspect terms associated with the opinions are listed in Table 2.3.

Much of the earlier work in sentiment analysis is confined to manual feature engineering tasks and using machine learning classifiers like NB, SVM, DT, ME, etc., on these features [8]. Salient features can be captured for this difficult task through deep learning which has proved to be successful in classification and CNN has given remarkable results. Deep learning can automatically identify features in the sentences. In this work, the implementation of the deep learning model, i.e., CNN is used for sentiment classification and its performance is compared with traditional machine learning algorithms on different datasets. As explained in OSMS in Fig. 2.3, the filtered opinions are stored in the pre-processed opinion repository and aspects with its adjectives are stored in the opinion word repository. The user can obtain the classified opinions on the basis of aspects by consulting the opinion word repository. For example, positive and negative opinions can be found in the classified opinion repository on the aspect 'battery' for the query posted as 'iPhone'. 'Opinion Classification' works on the algorithms of deep learning and uses a single layer of convolutional neural network to give the classified results in the positive or negative classes. The cleaned opinions [9] stored in the filtered opinion repository with tokenized words are mapped into Word2Vec embedding and fed into the layers of CNN. The classified opinions are stored into the classified opinion repository which serves as the response to the user's query. The details and the analysis of the various classifiers along with CNN model are elaborated in Chap. 4.

2.2.3 Aspect Based Opinion Summarization

Summarizing the opinions involves the use of two repositories, one for extracting the relevant aspects of the opinion word repository and the other is classified opinions repository. The summarization is based on constructing graphs for sentences and ensuring its correct based on natural language rules. The sentences are then scored and ranked using an extractive technique by implementing PCA on the matrix created using SentiWordNet score. The results are evaluated using ROUGE tool and detailed comparison of both techniques is discussed in graphs and charts which is explained in detail in Chap. 5.

2.2.4 Look up for Alternate Data

Some alternate data sources are used by OSMS to respond the user's query if it is not able to provide the satisfactory response to the user. If the user is satisfied with the response given by alternate data sources, then opinion retriever updates its *URL Database* with the data obtained from the alternate data sources. If the user is not satisfied with the answer given by the alternate data source, then the answer is simply discarded and OSMS does not update its database. Also, the OSMS displays an appropriate message in this case. The algorithm for *Look up for alternate data sources* modules is given in Fig. 2.4.

The updating of the *URL Database*, with a satisfactory answer(s) returned by the alternate data source is a major step taken for enriching the database maintained by the proposed system. It may be worth noting that this step takes place in increments and enriches the OSMS database gradually.

The flowchart of the whole process is shown in Fig. 2.5.

Algorithm Look up for alternate data sources ()

{

      POST the query from the alternate data sources

      Compile the response(s) from the response page obtained and present the response(s)

      to the user on OSMS interface;

      if (the user is satisfied with the response given)

            Update the URL database of Opinion Retriever with the response page obtained as

            new seed URLs;

      else

            display a message" sorry, no summarized opinion for your query"

}

Fig. 2.4 Algorithm: look up for alternate data sources

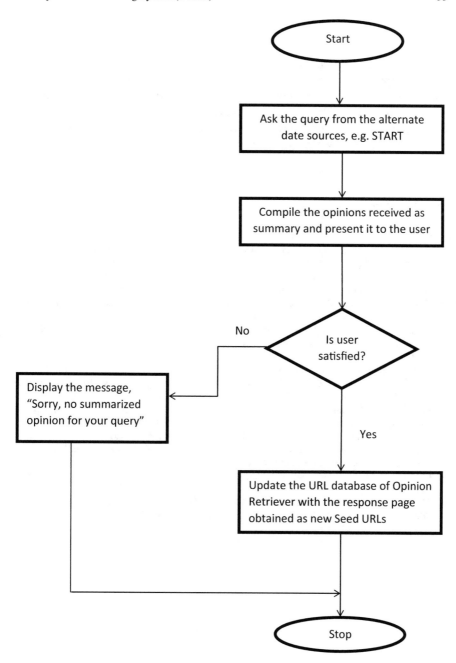

Fig. 2.5 Flowchart: process for alternate data sources

2.3 Summary

The solution is proposed which is used by end users to make quick decisions by revealing opinions on a particular topic as mentioned by the user query; the results are returned as facts and opinions specific to that domain; ignoring the opinions not related to the domain. OSMS details out the algorithms and techniques for identifying aspects, classifying opinions, and summarizing opinions. The proposed OSMS includes several steps which are explained in detail in the following chapters.

References

1. Ku, L. W., Liang, Y. T., & Chen, H. H. (2006, March). Opinion extraction, summarization and tracking in news and blog corpora. In *AAAI Spring Symposium: Computational Approaches to Analyzing weblogs* (Vol. 100107).
2. Madaan, R., Sharma, A. K., & Dixit, A. (2012, December). A novel architecture for a blog crawler. In *Parallel Distributed and Grid Computing (PDGC), 2012 2nd*.
3. Uysal, A. K., & Gunal, S. (2014). The impact of preprocessing on text classification. *Information Processing and Management, 50*(1), 104–112.
4. Collobert, R., Weston, J., Bottou, L., Karlen, M., Kavukcuoglu, K., & Kuksa, P. (2011). Natural language processing (almost) from scratch. *Journal of Machine Learning Research, 12*(Aug), 2493–2537.
5. Naik, A., Satapathy, S. C., Ashour, A. S., & Dey, N. (2018). Social group optimization for global optimization of multimodal functions and data clustering problems. *Neural Computing and Applications, 30*(1), 271–287.
6. Bhatia, S., & Madaan, R., (2019). An algorithmic approach based on PCA for aspect based opinion summarization. Paper presented in Special Session (9.004) of INDIACom-2019. IEEE.
7. Ali, M. N. Y., Sarowar, M. G., Rahman, M. L., Chaki, J., Dey, N., & Tavares, J. M. R. (2019). Adam deep learning with SOM for human sentiment classification. *International Journal of Ambient Computing and Intelligence (IJACI), 10*(3), 92–116.
8. Liu, B. (2015). *Sentiment analysis: Mining opinions, sentiments, and emotions.* Cambridge University Press.
9. Bhatia, S., Sharma, M., &Bhatia, K. K. (2016). A novel approach for crawling the opinions from World Wide Web. *International Journal of Information Retrieval Research (IJIRR), 6*(2), 1–23. (ESCI, WEB OF SCIENCE).

Chapter 3
Opinion Retrieval

3.1 Introduction

The massive amount of text is available online because of easy availability of Web technologies. This motivates users to post opinions and share their experiences on Web for decision-making process. Information can be searched easily on any topic using facts but searching opinions is difficult. Previous research and extraction systems are thoroughly analyzed for constructing an efficient retrieval system for opinion mining which will distinguish between constructive opinionated information. The research focuses on building 'opinion retriever' which will work as an extended layer on the existing search engine, i.e., Google. The revisit time of each URL is calculated using novel algorithm and Web pages that need to be revisited for collecting fresh opinions are crawled. The reviews which are extracted using opinion retriever undergo pre-processing tasks also. The activities involve cleaning operations to search for relevant content from opinion repository. The effectiveness is measured by considering different datasets.

The remainder of the chapter is organized as follows: Sect. 3.1 presents detailed introduction and the importance of extracting opinions from text and related work framework and algorithms, experimental evaluation, and results of opinion retriever. Section 3.2 discusses the meaning of opinion spam detection and its methods. Section 3.3 describes the definition of preprocessing, its various tasks, and the structural model proposed for cleaning opinions. Section 3.4 presents the crawling model and the algorithms associated with crawling. Section 3.5 concludes the overall proposed work of retrieving opinions.

© The Author(s), under exclusive license to Springer Nature Singapore Pte Ltd. 2020
S. Bhatia et al., *Opinion Mining in Information Retrieval*,
SpringerBriefs in Computational Intelligence,
https://doi.org/10.1007/978-981-15-5043-0_3

3.2   Extraction of Opinions from Text

Web pages are traversed and index of the pages are created with the help of bots
known as Web crawlers [1]. The search engine is processed with the basic goal of
indexing the downloaded HTML pages for fast searching by storing a copy of the
visited URL. It is given in Fig. 3.1.

The components explained below:

- URL Frontier: This contains the URLs as seed set and a crawler picks URL and
  starts its process.
- Domain name service resolution: The IP address is searched for the domain names.
- Fetch: The URL is picked and Hypertext Transfer Protocol (HTTP) is used for
  fetching it.
- Parse: This is concerned with analyzing the structure of the HTML code and
  includes parsing of the page. During this process, all the texts and images are
  extracted and links are extracted from the Windows Internet (WinInet) library
  using breadth first manner or depth first manner.
- Content Seen?: This involves checking of the Web page with the content. If it is
  already there within another URL, then alternative way is to find to measure the
  fingerprint of a Web page.

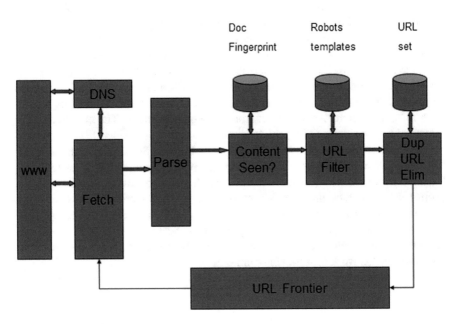

Fig. 3.1 Web crawler

3.2.1 Web Search Versus Opinion Search

Searching opinions from Web is a challenging task as Web is huge and quite challenging. The size of Web approximately is given in studies [2, 3]. There are billions of pages available in Web. Considering 10–15 KB approximate average size of a Web page, there is at least tens of terabyte textual data amount available on the World Wide Web. The inclination toward considering reviews and feedbacks by users on WWW for decision making has exponentially increased and large amount of unstructured user generated content has been gathered over the Internet with time. On posting a query, numerous Web documents relevant to the particular area are retrieved which are answered in the form of Web pages, but finding opinions from these Web documents is again another challenging task, thereby the most relevant and the recent ones. A lot of manual efforts will be required with huge amount of valuable time as searching for the opinion in that case would require navigating through different Web pages in detail.

3.2.2 Challenges in Retrieving Opinion

WWW is considered as an important source of collecting opinions which serve as a medium of setting trend of taking decisions based on the reviews and feedbacks by the users. The Internet is filled up with the unstructured data created by users. The opinion shared and communicated on social media platforms persuade the decision making and policy making of the organization's leaders for the product. The cumbersome task of summarizing and extracting meaningful and appropriate information needs data mining techniques which makes the task easy. The authors of [4–6] show their work on analyzing reviews and mining opinions. Opinion mining aims to provide preference level illustration of the opinions to users, main features of opinions at the document level, and precise abstract at sentence level. An efficient Web crawler can extract this valuable information from a variety of web sites like review/survey sites, E-commerce, and social media web sites. Crawling traverses the Web documents and gathers the information over the text Web pages [1, 7, 8]. The reviews and feedback from users constantly change over time on some parts of the Web pages. The work of crawler is to revisit, search, and retrieve the updated reviews and feedbacks posted on these dynamically changing data-driven Web pages [9]. Thus, main focus here is to calculate the frequency of revisit of crawler in context of collecting opinions from these Web pages.

3.2.3   Existing Opinion Retrieval Techniques

There has been immense work done on extracting reviews by various researchers. However, to build the retriever that crawl opinions and updates its opinions on temporal basis is still a task [10, 11]. The proposed work in this work deals with developing the opinion retriever which updates it timely is discussed in the next section. Some of the related work done by the prominent researchers and the issues in this field are discussed below.

The authors [12] have designed a heuristic algorithm EffMaxCover to extract the relevant reviews. The review selection problem is entified and a new technique is presented that aims to find a minor set of reviews that properly cover the microreviews. The matching of reviews is done syntactically and semantically; the algorithm is then applied on the pairs to view the output. The experiments have been conducted on the corpus collected for restaurant reviews and results are evaluated. The paper [13] has proposed a decisive predictive model with two-level filtering approaches for extracting reviews online. The reviews are scored, filtered, and a sentiment score associated with each feature of the product is calculated for mining the reviews. Although, the deviation is calculated for both subjectivity probability and feature score for each review manually and through the proposed methodology, still the accuracy needs to be improved. The issue of word-1sense disambiguation needs to be rectified. The authors [14] have presented a novel technique of assessing product reviews. The algorithm works in two steps. First, the graph-theoretic clustering methods are used to divide features into clusters. Secondly, the most representative feature is chosen from each cluster to form a subset of features. The features are independent. The researchers [15] have proposed a semantic-based product feature extraction technique. The summary of opinions about each extracted product feature is generated using different combinations of typed dependencies. Improvement can be made by combining hybrid rules using typed dependencies. The authors [16] have implemented the distributed Web crawler in which multi-threaded approach is followed. The paper discusses the overall architecture and the problems that occur during implementation with their solutions with DC crawler. The overall performance is tested with large datasets which leads to be more effective than traditional crawlers. The paper [17] has discussed the focused crawler which extracts the Web pages that are relevant according to the particular topic. It also verifies the text by applying pattern search algorithm. The authors [18] have analyzed the real-time data and developed an online system for retrieving opinions topic wise on the Web with the complete framework illustrated with case studies. The two main contributions are focused, crawler and miner agents which produce genre aware results of opinion retrieval. The paper [19] has proposed a tool to collect and visualize data from face book. It has discussed the overall system architecture which can extract opinions from social networking sites on the particular topic. The implementation is discussed and testing is performed in Java. The tool developed can perform frequency analysis. The researchers [20] have developed a novel focused based crawler specific to the topic. The page relevance is predicted by the classifier by following content-based approach for building Web comments collections. The implementation and evaluation of the proposed system prove to be more effective in terms of precision and recall rates of Web comments. The authors [21] have designed a crawler based on Ajax technology in microblog. The methodology was based on event-driven collections to extract data maintaining the efficiency of obtaining information and integrity. The paper [22] has proposed a meta-crawler framework which supports the search engines suggested by the user. The clustered results retrieved in the form of Web pages enable users to find their groups easily. The proposed algorithm also serves better in terms of efficiency and performance as it resolves the problem of topic drift.

Web is a huge source of unstructured data. Web data extraction systems are related interactively with the Web sources which extract the content stored in it [23]. But, there are certain situations that may pose a large number of problems which need to be handled carefully while extracting data. Some of the problems are unpredictable growth of data on the Web, high degree of semantic heterogeneity [24], privacy concerns, formats often tend to change and improper schema, data integrity and maintenance, temporal dependency.

The following limitations have been discussed in converse literature:

- The wrapper approach used by feature extractor mostly based on template which enhance the maintenance cost of the template automatically.
- The scalability has been compromised by current approaches due to complex algorithms.
- The extraction methods discussed in literature do not evolve with respect to time, thus scope of these methods is limited.
- The updating and changing behavior of Web pages need retraining of proposed machine learning algorithms.
- The quality of local collection is not handled properly by discussed techniques in literature.
- The discussed heuristic approaches also produce irrelevant content/information which is no use of users/topic.
- The application specific wrapper models built previously are not generalized in way to spread to broader domain related to E-commerce.

All of these disadvantages of the earlier work are considered and solutions have been proposed in the later sections in this chapter.

3.3   Opinion Spam Detection

Opinion mining is becoming eminent field used by organizations and individuals for taking decision for purchase or bigger business decisions. This also increases the fake opinion creation to promote particular product and demote other competitor's products. For example, an individual can write a good review about a good product and can use it for a bad product in order to false promotion of other bad product. These fake opinions which deceive the customers are usually called opinion spam or opinion spamming or deceptive reviews. The individuals who create these fake reviews can be called as opinion spammers. Opinion spamming becomes more frightening when spammers create fake reviews about political and social issues. Therefore, opinion spam detection methods have attracted many researchers and organizations to detect the spam opinions as early as possible to avoid the impact. Jindal and Liu [25] first published their research on quality control on these deceptive opinions which can be written by anyone using Web services. They compared the opinion spam as similar as Web page spam which illegitimately boosts the ranks of particular Web page to come at top in search engine result. Web spam furthers can be divided into link spam and content spam. But the opinion spam is much different from Web spam. Unlikely to link spam reviews, hardly contains the hyperlinks. Also unsolicited advertisement and irrelevant words are rarely present in the reviews posted. The authors [25] used the product reviews from Amazon.com in the absence of gold standard opinions and trained their models on the extracted features like reviewer, product and review content to detect the deceptive spam and truthful reviews. An alternate approach based on positive deceptive opinion spam Ott et al. [26] estimated the 1 and 6% of positive opinion for hotels on a travelling web site are deceptive and posted by the hotels for fake promotion. A negative deceptive opinion spam has explained by Ott et al. [27]. They have used 400 gold standard negative reviews of 20 hotels of USA and performed the analysis manually by three volunteer judges and two meta-judges to evaluate negative deceptive opinion and to improve the performance by using linear support vector machine.

The foremost challenge in Opinion spam detection is to recognize fake reviews by manually reading them and use it to devise the general detection algorithms. It has been comparatively tough to detect spam in opinion mining than other form of spam.

3.3.1 Spam Types

There are following three types of opinion spam [25].

Type 1 [Fake Opinions]: Spammers deliberately write fake reviews to promote or defame the target product and services. They can either be positive fake reviews to falsely promote the bad product/services or negative fake reviews to defame the competitive product/services.

Type 2 [Biased Opinions]: The reviews which are biased due to the brand of products and services are biased opinions. It contains the review about the brand despite of quality of the product and services. For example a review for Apple smart phone says '*This phone doesn't have battery backup. Do not buy it.*'

Type 3 [Non-reviews]: Irrelevant information other than opinion falls under this category. This can be further divided in two parts: 1. Advertisements and 2. Text-like questions, random texts, answers.

Type 1 is more damaging than Type 2 and Type 3. The authenticity of these reviews is unpredictable by manually reading. While Type 2 and Type 3 can be easily be detected by manual reading.

3.3.2 Fake Reviews

Fake reviews are difficult to detect and studied as special cases by many researchers [26–35]. Basically, fake reviews are about detecting individual's true feelings or lies about the facts. Deceptive intentions in text have been predicted by the researchers. For example, study shows that news containing facts has different content style than fantasy or fake news. Quantity of nouns, verb, noun phrases, paragraph count, and repetition of words are some style attributes to detect the news is factual information or fake information. Fake opinion significantly differs from lies each other. Many psychological theories say that when a person lies, it avoids to use words like I, my, myself, etc. But spammers use these words frequently. For example, an application developer can pretend to be a reviewer and write the positive review about the app. So the feeling of the developer about his/her application will contain the words like I, my, etc., also it's possible that reviewer have not used the services or product but writing a bad review without knowing the facts.

Fake reviews can be extremely harmful in special cases like political or social topics, new launched product or services. Opinion spammer can be of any categories like employees, friends, business organization, family, actual consumer, professional spammers hired by competitors, etc. The meta-data about review, behavioral patterns, and product information can be also helpful to detect Type 2 and Type 3 spams. Meta-data of review contains rating of the product, reviewer's location, IP address, MAC address, etc. The data used to produce fake reviews can be distinguished as public data like location of reviewer, number of repetitive reviews, etc., and private data like Mac address, IP address, cookies, etc.

3.3.3 Spam Detection Methods

Opinion spam has significant influence on opinions of users. It can be harmful and created by any spammer. As discussed earlier, Type 2 and Type 3 spamming can be easily detected manually using discussed attributes. The main topic of discussion is 'how to detect and isolate fake reviews, how to detect fake reviewers and how to find fake reviewer groups.' All answers of these three questions are correlated in different aspects. So spammer, spam review, and spammer group should be detected as soon as possible by exploiting their own characteristics.

Two strategies for spam opinion detection have been discussed in this section, i.e., supervised spam detection and unsupervised spam detection.

3.3.3.1 Supervised Spam Detection

As discussed earlier, it is very hard to recognize and predict the fake reviews by reading manually. The spammers write the reviews very carefully to make it equivalent to actual review. So, the reliable dataset for training machine learning algorithms for fake and non-fake reviews are not available. Despite of these difficulties, many researchers have proposed and evaluated machine learning algorithms and formulated the opinion spam detection problem into classification problem of two classes, i.e., *fake class* and *non-fake class*. Three supervised learning methods have included in this section.

The authors [26] exploited the duplicate reviews in their study. They took 2.14 million reviewers, 5.8 million reviews, and 6.7 million products from *Amazon.com* and concluded their study with large number of duplicated reviews by same reviewers or different reviewer. They used shingle method [36] for finding duplicate/not-duplicate review and used it to train their model using logistic regression. The features were divided into three categories as follows: (1) review centric features, (2) reviewer centric features, and (3) product centric features.

Li et al. [37] used a manually labeled corpus of epinions reviews and trained their model based on it. Supervised machine learning algorithms like SVM, logistic regression, Naïve Bayes have been applied on manually created corpus and found significant results. Ren and Ji [38] have used neural network for deceptive spam detection. Ott et al. [26] have used TripAdvisor for truthful reviews to find fake reviews from Amazon Mechanical Turk. They have constructed feature set on the basis of different categories as follows: (1) Genre identification, (2) Psycholinguistic deception detection, (3) Text categorization and (4) Human/Meta data. Naïve Bayes and SVM have been trained with these features set and resulted that the reviews were from unseen hotels. Many more supervised researches can be seen in [34–38]. Despite of these researches, there is still uncertainty in reliability of whether the review is fake or non-fake. This creates a scope of study in this field.

3.3.3.2 Unsupervised Spam Detection

Unsupervised learning-based methods for opinion spam detection can be helpful as manual labeling if training data is difficult.

Generative Model: Two types of cluster can be formed as fake cluster and non-fake cluster using unsupervised Bayesian clustering. It takes the fakeness of reviews/reviewers as latent variables and the latent variable-based clustering falls into the generative models for clustering [39]. The behavioral pattern features like content similarity, reviewing activity, extreme rating, rating deviation, etc., can be observed for finding fake/non-fake reviews.

Atypical Behavior-based Model: These models are based on a typical behavioral patterns present in the reviews. For example, if a reviewer writes positive review about a T.V. while all other reviewer write negative review about the same T.V., then former reviewer can be considered as suspicious. Authors [31] have suggested a technique to find unusual reviewer behavior based on patterns by assigning a numeric score. The score is assigned on the extent of fake practices of the reviewer and then this score summed up to produce final spam score of the reviewer.

All these methods use heuristics features to find the spam/non-spam reviews. Class association rules are domain independent generic method and use the data mining class association rules for finding spam opinions. A class of association rules has to be created by creating a class and dataset of attributes like reviewer id, product, brand, etc., for each review. Class association rules are considered as association rule mining.

3.4 Cleaning Opinions

The detailed structure of cleaning text for opinions is given in Fig. 3.2. It consists of mainly three steps which are explained as follows:

- Opinion extractor
- Special symbols, Stop word removal, Lemmatization (SSL)
- Tokenizer.

All the steps are explained below:

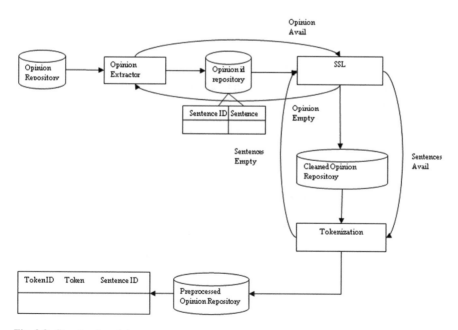

Fig. 3.2 Structural model

OPINION EXTRACTOR:

Stanford Core NLP library version 3.4 annotators are used to split sentences, which allow working at a sentence level. By splitting sentences, we can draw the boundaries, which in turn let us continue working under the assumption that the aspects and its corresponding opinion can be found within the sentence boundaries. This is how the dataset appears after it has drawn the sentence boundaries:

- Do not buy this piece of junk.
- The phone has bad screen.
- I find it uninteresting.

This first module will extract all the opinions from the opinion repository and will break the opinions into individual sentences with their particular sentence ID. The sentences will be stored in the opinion ID repository. This will further send the opinion avail signal to the SSL to clean the sentences. If no sentences are available in the opinion ID repository, then SSL will send opinion empty signal to the opinion extractor. The procedure is given in Fig. 3.3.

SSL

This constitutes three tasks to achieve: Special symbols removal, stop word removal, and lemmatization. All these tasks (shown in Fig. 4.14) are performed and after cleaning the sentences, the sentence is stored in cleaned opinion repository. The procedure is shown in Fig. 3.4.

TOKENIZER:

Tokenization will make our review sentences to split in tokens. Stanford POS tagging tool[1] [40] has been used for POS tagging of the documents, which tag all the documents and then generates an output file as POS-tagged XML document [41, 42]. The tokens will be stored in the pre-processed opinion repository. Tokenizer will send the *SentenceEmpty signal back* to SSL if no sentences are available. The procedure is shown in Fig. 3.5.

```
Opinion Extractor()
{
wait(Opinion Avail)
fetch Opinions from the Opinion Repository
break Opinions into individual sentences with
Sentence ID
Store in Opinion ID Repository
signal (Opinion Empty)
}
```

Fig. 3.3 Opinion extractor

```
SSL()
{
wait(Sentence Avail)
Fetch sentences from the Opinion id Repository
Remove special symbols
Apply stop word removal
Apply Lemmatization
Store in Cleaned Opinion Repository
Signal (Sentence Empty)
}
```

Fig. 3.4 SSL

```
Tokenizer()
{
wait(Sentence Avail)
Fetch Sentences from the Opinion ID Repository
Tokenize the sentences into tokens
Store in Pre-processed Opinion Repository
Signal (Sentence Empty)
}
```

Fig. 3.5 Tokenizer

Finally pre-processed opinion repository will contain the filtered tokens with POS associated with it, token ID, and sentence ID. For example, the first opinion listed in Table 3.2 is 'The room was packed to capacity with queues at the food buffets.' The cleaning tasks performed on the above opinion will represent in the pre-processed opinion repository as shown in Table 3.1.

3.4.1 Preprocessing and Its Tasks

The following are the pre-processing tasks [43] applied to the opinions. The pre-processing is accomplished using NLP techniques.

- Removal of special symbols: The dataset is cleaned using regular expressions, where symbols such as {, [,:),:(…, are removed since the reviews are natural text. The reviews given in Table 3.2 are taken as examples.
  - '[t] do not buy this piece of junk.'
  - '## The phone has bad screen. I find it uninteresting.'

Once the dataset has been cleaned out, this is how it appears:

Table 3.1 Example of pre-processed opinion

Token ID	Token	Sentence ID
1	Do/VBP	1
2	Not/RB	1
3	Buy/VB	1
4	Piece|NN	1
5	Junk|NN	1
1	Phone/NN	2
2	Bad/JJ	2
3	Screen/NN	2
1	I/PRP	2
2	Find/VBP	2
3	Uninteresting/JJ	3

Table 3.2 Opinions stored in the opinion buffer

Opinion ID	Opinion	$D_{opinion}$	$T_{opinion}$
1	[t] do not buy this piece of junk	April 18, 2015	7:10
2	## The phone has bad screen. I find it uninteresting	April 19, 2015	21:00
3	Perfect nothing scratched. One of them with wifi problem, need to replace to improve the wireless	April 19, 2015	21:30
4	Very disappointed bought this phone this year and now the battery has spoil guess because the phone did not came with the original charger	April 20, 2015	21:30
5	Damaged product. Very upset with the product. Product arrived in good condition but battery was not working.. I've had to replaced battery by own	April 21, 2015	22:10
6	Worked good. The phone was in good condition. Looked good as new. Only one problem, the camera sound does not get silent. The phone volume gets silent, just the phone wouldn't get silent while clicking the pictures	April 21, 2015	6:20

- (do not buy this piece of junk.)
- (The phone has bad screen. I find it uninteresting.).

- **Stop word elimination**: It is a process of eliminating the commonly occurring or rarely occurring words existing in a sentence like a, an, the, are, etc. The process of removing the stop word is an important data preprocessing task. Stop words are language dependent functional words and do not carry any relevant information of the topic [44]. Removal of these words leads to removal of unwanted words in each review sentence. The stop word list contains words like preposition, pronouns, conjunctions, etc.

Fig. 3.6 Output of Stanford POS tagger

- Lemmatization: The dictionary form of a word is returned by making use of a vocabulary. This is known as removing inflections called as 'lemma.' For instance, the word 'walking' will return 'w' with stemming but 'walk' or 'walking' with lemmatization depending on whether the use of the token was as a verb or a noun. Stanford lemmatization removes the endings of words and returns the word to the base or dictionary form of the word. It then allows us to form single item by grouping different, forms of words together.
- Tokenization: After this step, the dataset is ready for the next step (POS tagging). Breaking the sentences into fragments called tokens is the work of tokenized which also removes certain special characters such as punctuation. Tokenization will make review sentences to split in tokens [40, 44]. The tagged sentence is represented in Fig. 3.6.

3.5 Crawling Opinions

The seed URLs used by a general crawler for collection of Web pages can be added on as the process works. The new Web page searched by the crawler in this process is then added to the list of queue. The explained process repeats until all the URLs covered in the queue list or crawler is manually interrupted. This collects the abundant amount of text as Web pages. This is mandatory to extract/download these abundant Web pages in order to fulfill the goal of opinion mining. This paper purposes opinion extractor designs which not only downloads the Web pages and extract the opinions of various products of the users but also keep the repository updated with latest/refresh opinions by incremented crawling over Web. The detail of the proposed design of the opinion retriever/extractor is explained below.

The downloaded seed URLs are extracted into the URL database. The seed URLs along with precedence associated are stored in the form of the queue. The calculation of associated precedence is done by calculating the average time difference of the existing opinions in that URL and the new opinions posted. The URL has the highest probability of having new opinions if it has the lowest precedence. So the precedence is inversely related to the probability of new opinions. The precedence's of *SeedURLs* are given as 0. The URL extractor extracts the URLs from the database and sends the *URLAvail* signal to the opinion follower. The opinion follower fetches the URLs from *URLRegister* and checks if new opinions are available by searching the div tag (URL) or span tag (URL). It then searches for the review text, comments, opinions/summary in the extracted HTML code. It sends the *OpinionURLAvail* signal to the opinion downloader after validating the URL. The work of opinion downloader is to extract and download the corresponding Web pages from WWW of URLs which are accumulated in the *OpinionBuffer*. It further stores these Web pages in the Page Repository. Otherwise, opinion downloader sends the *OpinionRegisterEmpty* signal to the opinion follower. Next, the opinion extractor analyzes the structure of HTML page and extracts the opinions from the Web page by using the string matching technique. The opinion buffer contains the extracted opinions with the link information like time and date of the opinion posted on the Web as text. The work of the revise module is further extraction of the URLs which are stored individually in a file and calculations of the precedence of the each individual URL. It stores the calculated precedence in URL database. The URL extractor then fetches the lowest precedence URL and the process continues. Figure 3.7 demonstrates the architecture of proposed design.

- URL extractor
- Opinion follower
- Opinion downloader
- Opinion extractor
- Opinion buffer
- Revise.

The procedure for URL extractor is given in Fig. 3.8.

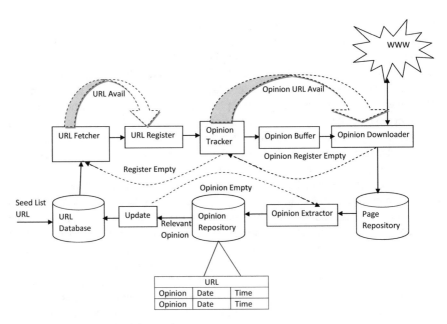

Fig. 3.7 Architecture of opinion retriever

```
URL Extractor()

{
wait(RegisterEmpty);
if (precedence==0)
extract URL from the URL Database;
else
{
compare the precendence of the URLs;
extract URL with the lowest precedence;
}
store it in URLRegister;
signal(URLAvail);
}
```

Fig. 3.8 URL fetcher

The procedure for opinion follower is given in Fig. 3.9.

Figure 3.10 shows the example taken from one of the review sites.

The procedure for opinion downloader is given in Fig. 3.11.

Opinion extractor will store the opinions in the opinion buffer in the text file with the following information.

$D_{opinion}$: date of the opinion

$T_{opinion}$: time of the opinion

This is shown in Fig. 3.12.

Table 3.2 shows how the opinions are stored in the opinion buffer.

```
Opinion Follower()

{
wait(URLAvail);
fetch URL from URLRegister;
if(span tag(URL) or div tag(URL))
{
if("reviewtext(URL)"||"opinion(URL)"||"com
ment(URL)"||"summary(URL)")
{
store in OpinionBuffer;
signal(URLOpinionAvail);
}
else
signal(RegisterEmpty);
}
```

Fig. 3.9 Opinion tracker

Span tag

Review
text

Fig. 3.10   Snapshot of the HTML code

```
Opinion Downloader()

{
wait(OpinionURLAvail);
extractOpinionURL from OpinionURLBUffer.
download web pages corresponding to the
respective URL;
store in PageRepository;
signal(OpinionBuffEmpty);
}
```

Fig. 3.11   Opinion downloader

$D_{opinion}$

$T_{opinion}$

Fig. 3.12   Snapshot of the opinion

There is an important role of the module revise as it will revise the opinion buffer with the latest and fresh opinions by using the formula.

$D_{opinion}$: date
$T_{opinion}$: time.
$D_{new\ opinion}$: date
$T_{new\ opinion}$: time

The above-mentioned information is further used for calculating the precedence of each URL.

Revisit frequency can be computed by finding out the appropriate rate at which the retriever will revisit the URL again [7]. Update module reads the URL's in sequence and the frequency is computed using the formula given below:

$$\Delta t = D_{new\ opinion} - D_{opinion} + T_{new\ opinion} - T_{opinion} \tag{3.1}$$

$$\Delta t_{avg} = \sum_{i=2}^{n} \Delta t_i / n - 1 \tag{3.2}$$

$$\Delta t_{avg} = (\Delta t_2 + \Delta t_3 + \Delta t_4 + \Delta t_5)/4$$
$$= 206,400/4$$
$$= 51,600\,s$$
$$= 0.59722222\,days$$

The procedure for the revise module is given in Fig. 3.13.

Revise ( )
{
URl's fetched with related information from Page Repository;
for each URL$_i$ in Opinion buffer for n opinions
{
compute average time difference
$\Delta t_{avg} = \sum_{i=2}^{n} \Delta t_i / n-1$;
precedence=Δt_{avg}
add URL and its precedence in URL Database
}
signal(*OpinionEmpty*);
}

Fig. 3.13 Revise module

The work proposed above proves the retriever to be incremental as it incrementally refreshes the existing collection of pages by visiting them frequently according to the precedence computed as above [45].

3.6   Summary

The various algorithms and techniques are proposed that help to crawl the Web pages by presenting the recent and fresh opinions to the user. The process of computing the revisit frequency for each URL is explained. The chapter explains the different algorithms that can be used to find the updated and relevant opinions. The comparison with the previous work already done in this area proves the effectiveness of the proposed work. The effort for constituting the overall OSMS system relies on extracting and cleaning the opinions for getting the final summarized opinions categorized according to the features.

References

1. Arasu, A., Cho, J., Garcia-Molina, H., Paepcke, A., & Raghavan, S. (2001). Searching the web. *ACM Transactions on Internet Technology (TOIT)*, *1*(1), 2–43.
2. Lawrence, S., & Lee Giles, C. Accessibility of information on the web. *Nature*, 400:107–109.
3. Bar-Yossef, Z., Berg, A., Chien, S., Fakcharoenphol, J., & Weitz, D. (2000). Approximating aggregate queries about web pages via random walks. In *Proceedings of the Twenty-sixth international conference on very large databases*.
4. Pang, B., Lee, L., & Vaithyanathan, S. (2002). Thumbs up?: sentiment classification using machine learning techniques. In *Proceedings of the ACL-02 conference on empirical methods in natural language processing* (Vol. 10, pp. 79–86). Association for Computational Linguistics.
5. Hu, M., & Liu, B. (2004a). Mining opinion features in customer reviews. In *AAAI* (Vol. 4, No. 4, pp. 755–760).
6. Liu, B. (2012). Sentiment analysis and opinion mining. *Synthesis Lectures on Human Language Technologies*, *5*(1), 1–167.
7. Brin, S., & Page, L. (1998). The anatomy of a large-scale hyper textual Web search engine. *Computer Networks and ISDN Systems*, *30*(1–7), 107–117.
8. Burner, M. (1997). Crawling towards Eternity: Building an archive of the World Wide Web. *Web Techniques Magazine*, *2*(5).
9. Madaan, R., Sharma, A. K., Dixit, A. (2012). A novel architecture for a blog crawler. In *2012 2nd IEEE international conference on parallel, distributed and grid computing*. IEEE.
10. Dey, N., Wagh, S., Mahalle, P. N., & Pathan, M. S. (2019). *Applied machine learning for smart data analysis*. CRC Press.
11. Dey, N., Das, H., Naik, B., & Behera, H. S. (Eds.). 2019. *Big data analytics for intelligent healthcare management*. Academic Press.
12. Nguyen, T. -S., Lauw, H. W. & Tsaparas, P. (2013). *Using micro reviews to select an efficient set of reviews*. ACM 978-1-4503-2263-8/13/10, San Francisco, CA, USA[4]: CIKM'13.
13. Soundarya, V., Rupa, S. S., Khanna, S., Swathi, G., & Manjula, D. (2013). Extracting business intelligence from online product reviews. *International Journal on Soft Computing (IJSC)*, *4*(3)[5].

14. Song, Q., Ni, J., & Wang, G. (2013). A fast clustering-based feature subset selection algorithm for high-dimensional data. *IEEE Transactions on Knowledge and Data Engineering, 25*(1), 1–14.

15. Kumar, R., & Raghuvee, K. (2012). Web user opinion analysis for product features extraction and opinion summarization. *International Journal of Web & Semantic Technology (IJWesT), 3*(4), 6.

16. Shkapenyuk, V., & Suel, T. (2002). Design and implementation of a high-performance distributed web crawler. *ACM Digital Library*.

17. Gupta, P., & Johari, K. (2009). Implementation of web crawler. In *2009 2nd international conference on emerging trends in engineering and technology (ICETET)* (pp. 838–843). IEEE.

18. Pappas, N., Katsimras, G., & Stamatatos, E. (2013). Distinguishing the popularity between topics: A system for up-to date opinion retrieval and mining in the Web. www.inevent-project. eu/files/pappas.

19. Mfenyana, S. I., Moroosi, N., Thinyane, M., & Scott, S. M. (2013). Development of a Facebook crawler for opinion trend monitoring and analysis purposes: case study of government service delivery in Dwesa. *International Journal of Computer Applications, 79*(17).

20. Neunerdt, M., Niermann, M., Mathar, R., & Trevisan, B. (2013). Focused crawling for building web comment corpora. In *Consumer communications and networking conference (CCNC)* (pp. 685–688), IEEE, ISBN: 978-1-4673-3131-9, 11-14.

21. Xiaohong, Y., & Sisi, Z. (2012). Research and implementation of the technology supporting micro blog data collection based on web crawler. In *International conference on automatic control and artificial intelligence (ACAI)* (pp. 1674–1677). ISBN: 978-1-84919-537-9.

22. Babu, K. R. R., & Arya, A. P. (2012). Design of a metacrawler for web document retrieval. In *12th international conference on intelligent systems design and applications (ISDA)* (pp. 478–484). ISSN: 2164-7143, 27-29.

23. Ferraraa, E., Meob, P. D., Fiumarac, G., & Baumgartnerd, R. (2014). Web data extraction, applications and techniques: A survey. *Knowledge Based Systems, ACM.* arXiv:1207.0246v4 [cs.IR].

24. Kolkur, S., & Jayamalini, K. (2013). Web data extraction using tree structure algorithms— A comparison. *International Journal of Recent Technology and Engineering (IJRTE), 2*(3), 2277–3878.

25. Jindal, N., & Liu, B. (2008). Opinion spam and analysis. In *Proceedings of the 2008 international conference on web search and data mining.* ACM.

26. Ott, M., et al. (2011). Finding deceptive opinion spam by any stretch of the imagination. In *Proceedings of the 49th annual meeting of the association for computational linguistics: Human language technologies-volume 1.* Association for Computational Linguistics.

27. Ott, M., Cardie, C., & Hancock, J. T. (2013). Negative deceptive opinion spam. In: *Proceedings of the 2013 conference of the north american chapter of the association for computational linguistics: human language technologies.*

28. Rubin, V. L., Chen, Y., & Conroy, N. J. (2015). Deception detection for news: three types of fakes. In *Proceedings of the 78th ASIS&T annual meeting: Information science with impact: Research in and for the community.* American Society for Information Science.

29. Pérez-Rosas, V., et al. (2017). Automatic detection of fake news. arXiv preprint arXiv:1708. 07104.

30. Lin, Y., et al. (2014). Towards online anti-opinion spam: Spotting fake reviews from the review sequence. In *2014 IEEE/ACM international conference on advances in social networks analysis and mining (ASONAM 2014).* IEEE.

31. Jindal, N., & Liu, B., Lim, E.- P. (2010). Finding unusual review patterns using unexpected rules. In: *Proceedings of the 19th ACM international conference on information and knowledge management.* ACM.

32. Heydari, A., et al. (2015). Detection of review spam: A survey. *Expert Systems with Applications, 42*(7), 3634–3642.

33. Mukherjee, A., Liu, B., & Glance, N. (2012). Spotting fake reviewer groups in consumer reviews. In *Proceedings of the 21st international conference on World Wide Web.* ACM.

34. Zhou, X., et al. (2019) Fake news: Fundamental theories, detection strategies and challenges. In *Proceedings of the twelfth ACM international conference on web search and data mining.* ACM.
35. Gong, Q., et al. (2018). DeepScan: Exploiting deep learning for malicious account detection in location-based social networks. *IEEE Communications Magazine* 56(11), 21–27.
36. Broder, A. Z. (1997). On the resemblance and containment of documents. In *Proceedings compression and complexity of SEQUENCES 1997 (Cat. No. 97TB100171).* IEEE.
37. Li, F. H., et al. (2011). Learning to identify review spam. In *Twenty-second international joint conference on artificial intelligence.*
38. Ren, Y., & Ji, D. (2017). Neural networks for deceptive opinion spam detection: An empirical study. *Information Sciences, 385,* 213–224.
39. Berkhin, P. (2004). *Survey of clustering data mining techniques, 2002.* San Jose, CA: Accrue Software.
40. Collobert, R., Weston, J., Bottou, L., Karlen, M., Kavukcuoglu, K., & Kuksa, P. (2011). Natural language processing (almost) from scratch. *Journal of Machine Learning Research*, *12*(Aug), 2493–2537.
41. Bhatia, S., Sharma, M., & Bhatia, K. K. (2015). Sentiment knowledge discovery using machine learning algorithms. *Journal of Network Communications and Emerging Technologies (JNCET), 5*(2), 8–12.
42. Lan, K., Wang, D. T., Fong, S., Liu, L. S., Wong, K. K., & Dey, N. (2018). A survey of data mining and deep learning in bioinformatics. *Journal of medical systems*, *42*(8), 139.
43. Uysal, A. K., & Gunal, S. (2014). The impact of preprocessing on text classification. *Information Processing & Management*, *50*(1), 104–112.
44. Jeyapriya, A., & Selvi, C. K. (2015). Extracting aspects and mining opinions in product reviews using supervised learning algorithm. In *2015 2nd international conference on electronics and communication systems (ICECS)* (pp. 548–552). IEEE.
45. Cho, J., & Garcia-Molina, H. (2003). The evolution of the web and implications for an incremental crawler. In *Proceedings of the 8th world wide web conference.*
46. Bhatia, S., Sharma, M., & Bhatia, K. K. (2017). Opinion score mining: An algorithmic approach. *International Journal of Intelligent Systems and Applications*, *10*(11), 34. (SCOPUS).

Chapter 4
Aspect Extraction

The three different classification levels of opinion classification as discussed earlier are: (1) aspect-level, (2) sentence-level, and (3) document-level. The aspect identification is the fundamental approach of aspect-based opinion mining, as aspects are those prime entities, which provide important information about an opinion. Many researchers have been extensively studying on this topic, and more than 55 techniques were recapitulated for the mining of explicit aspects [1]. Among all these approaches, syntactical method that combines rules between grammar dependency relations and target entities has been proved to be more successful; but still the task remains highly challenging because of the manner in which these rules are chosen. This chapter discusses the aspect identification problem by proposing a novel unsupervised method by combining theories of rational awareness with sentence dependency trees to identify aspects. The results are carried out on dissimilar datasets consisting of numerous opinions and comparison with previous rule-based approaches demonstrates the success of the work.

4.1 Product Features Mining

Thematic word is a term about which a review is expressed or which describe the theme of the review, for instance, 'this is a nice restaurant.' In this sentence, 'restaurant' is a thematic word about which sentiment is expressed. A novel approach is proposed that covers the complete domain area for thematic word extraction. Thematic words collectively represent main theme of the document. Thematic words are extracted on the basis of common-sense open-mind knowledge. Many researchers have worked to extract target entities from ConceptNet[1] [2, 3]. ConceptNet assertions are used to identify relevant concepts, and these concepts and synonyms of

---

[1]It is the largest common- sense repository consisting of more than 250,000 relations.

© The Author(s), under exclusive license to Springer Nature Singapore Pte Ltd. 2020
S. Bhatia et al., *Opinion Mining in Information Retrieval*,
SpringerBriefs in Computational Intelligence,
https://doi.org/10.1007/978-981-15-5043-0_4

the concepts are taken from WordNet [4]. These words are taken as nodes, which are used in construction of n-ary tree (where every node has at most n children) of specific domain. Many researchers have discussed and argued that the main motive of opinion mining is to examine and evaluate sentiments which are stated by people on WWW [5–7]. An opinion communicated and written on a particular feature such as in the sentence 'picture quality of the phone is very clear'; however, it is not easy to find out the entity on which the opinion is articulated, for example 'it is very wide.' NLP tools can provide the thematic words which are associated with the opinion words. It is distinctly showing that the feature is 'picture quality' which is connected with the opinion word 'clear.' However, in the second sentence, the feature screen can be deduced from the opinion word 'wide' using rules [8]. Further, for greater and almost complete coverage of all concepts, the present work works on extending the rule-based methodology by using grammatical relations between sentences to present the thematic-opinion word pair after filtering the meaningful terms present in the dataset relative to the particular domain. The syntactic relations are one-to-one correspondence, so, grammar dependencies will help in identifying thematic-opinion word pair more easily and effectively. In this chapter, a novel rule-based algorithm has been proposed that extracts the thematic words from the reviews. In order to prove novelty of the proposed approach, three methods are chosen as baselines and comparison is done on the basis of IR evaluation methods [9] between them.

4.2 Opinion Word Extraction

The work proposed is based on the use of binary relations (join pairs of words or phrases with the relationship between them) to identify features that can be categorized as positive or negative with respect to the orientation in which the opinions are expressed. An object, a part of object or an attribute, can be called as an aspect or thematic word or target entity, and opinion words are generally adjectives or adverbs (a modifier plus an adverb). Both target entity and opinion words are closely related to the opinion and are representatives of the opinion. Furthermore, 'words which contain the opinion' and 'target entities' have relations in opinionated expressions too. Thus, a dependency parser [10] can be used to identify these relations and then can be exploited for opinion extraction tasks. Taking collectively the entire thematic words (target entities) and the opinion words will facilitate the process of aspect identification and will help further in classifying sentiments. Figure 4.1 presents the general approach for aspect-based classification. The opinions collected via opinion retriever are processed via preprocessing tasks (as explained in the previous chapters) which are collectively taken for the extraction of aspects.

Then syntactic parsers are extracted using Stanford dependency Relations, which will map the dependencies between all words within the sentence in the form of relations (governor, dependent) [11, 12].

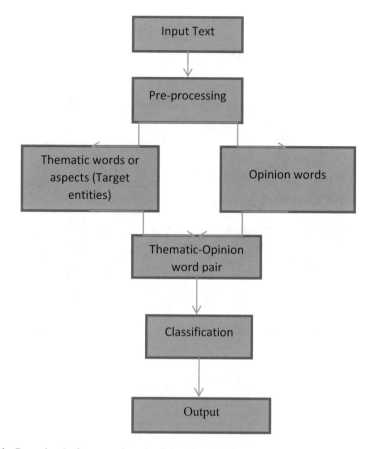

Fig. 4.1 General steps for aspect-based opinion classification

The Parse tree from the previous example discussed in Chap. 3 as 'do not buy this piece of junk' is illustrated in Fig. 4.2. The dependency trees are constructed for each sentence using Stanford Parser [13]. The first example as listed in Table 4.1 is considered.

After parsing opinions, the clear graphical view of optimal target entities and the optimal opinion words is presented [14, 15]. There are total 47 dependency relations listed in Stanford-typed dependencies manual; every sentence can trigger more than one dependency relation. In the proposed work, eight relations are used for defining rules to identify thematic-opinion word pair. Table 4.1 describes the relations with their abbreviations.

The work uses Stanford Parser (a natural language parser) for working out the grammatical structure of sentences. The phrases, each containing the thematic word and the associated opinion words, are identified from sentences. Detected phrases from sentences are mapped to aspects by using handcrafted dependency rules after construction of tree shown above. Some examples are shown for the above

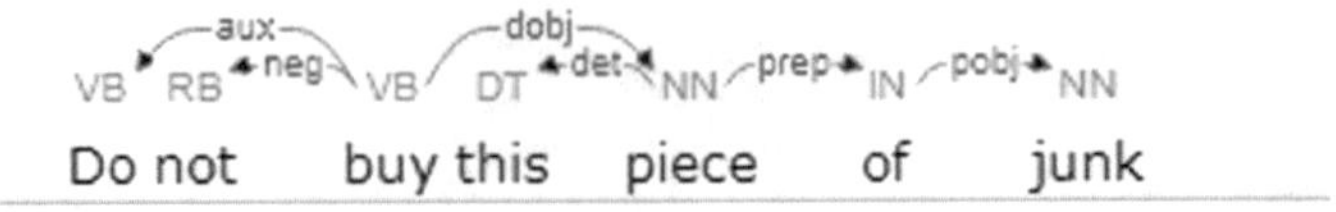

```
aux(buy-3, do-1)
neg(buy-3, not-2)
root(ROOT-0, buy-3)
det(piece-5, this-4)
dobj(buy-3, piece-5)
case(junk-7, of-6)
nmod:of(piece-5, junk-7)
```

Fig. 4.2 Dependency parser example

Table 4.1 Patterns

Typed dependency	Abbreviation
Adjectival complement	Acomp
Adjectival modifier	Amod
And conjunct	Conj and
Copula	Cop
Direct object	Dobj
Negation modifier	Neg
Noun compound modifier	Nn
Nominal subject	nsubj

relations used for aspect identification [16]; (as described in Stanford-typed dependencies manual). The aspects on the dataset Garmin Nuvi 255 W GPS are taken (listed in chap. 7).

The aspects are considered to be nouns and sentiment words to be adjectives, verbs, and adverbs, which have been widely used in previous works [17–19]. Therefore, dependency relations that frequently relate nouns (N and N), noun and adjective (N and J), noun and verb (N and V), and nouns and adverb (N and RB) are relevant for aspect extraction.

For example in the sentence 'The battery life is not long,' the term 'battery life' is related by the relation, compound (noun compounds), and both terms are noun.

Therefore, compound has a pattern of (N, N). Based on this condition, out of 47 dependency relations, 8 relations are used for experiments. The most useful dependency relations from previous work and the new dependency relations are listed after studying these rules in depth.

1. Adjectival complement (acomp): *The picture quality looks nice*, parsed to *acomp (nice, looks)*.
2. Adjectival modifier (amod): *This phone has great backup* parsed to *amod (backup, great)*;
3. 'And' conjunct (conj and): *This phone has great backup and resolution* parsed to *conjand(backup, resolution)*.
4. Copula (cop): *The screen is wide* parsed to *cop (wide, is)*.
5. Direct object (dobj): *I find it useful* parsed to *dobj (find, useful)*.
6. Negation modifier (neg): *The power backup is not prolonged* parsed to *neg (prolonged, not)*.
7. Noun compound modifier (nn): *The battery life is not long* parsed to *nn (life, battery)*.
8. Nominal subject (nsubj): *The screen is wide* parsed to *nsubj(wide, screen)*.

To extract more valuable opinion phrases, a hierarchical structure is built for these patterns, where N is noun, V is a verb, A an adjective, m a modifier, h a head term, and an opinion phrase $< h, m >$.

1. amod (N, Adj) â†' $< N, Adj >$
   This room has big space and view â†' (space, big)*

2. acomp (V, Adj) + nsubj(V, N) â†' $< N, Adj >$
   The room furniture looks beautiful –> (furniture, beautiful)

3. cop(Adj, V) + nsubj(Adj, N) â†' $< N, Adj >$
   The room is clean and spacious –> (room, clean)

4. dobj(V, N) + nsubj(V, N0) â†' $< N, V >$
   I hate the room view –> (room, hate)

5. $< t1, m >$ + conj and(t1, t2) â†' $< t2, m >$
   This room has big space and view –> (space, big), (view, big)

6. $< t, m1 >$ + conj and(m1, m2) â†' $< t, m2 >$
   The room is clean and spacious –> (room, clean), (room, spacious)

7. $< t, m >$ + neg(m, not) â†' $< t, not + m >$
   The room view is not happening –> (room view, not happening)

8.   < t, m > + nn(t, N) ↠ < N + t, m >
        The room view looks beautiful –> (room view, beautiful)

9.   < t, m > + nn(N, t) ↠ < t + N, m >
        I hate the room view –> (room view, hate)

4.3   Features Opinion Pair Generation

The proposed algorithm is given as follows:

Input = Opinion, Rules

Result = (Aspect,Opinion) list

For each sentence in opinion

   For each Rule 1 to 8:

     if sentence satisfies Rule i:

        aspect = getAspect(sentence)

        opinion = getOpinion(sentence

        Result.add(aspect,opinion)

Endif

EndFor

EndFor

Two examples given in Chap. 3 are taken, and the above algorithm is applied. The first example:

do not buy this piece of junk
Neg: (buy-3, not-2)
Dobj: (buy-3, piece-5)
Nmod: of (piece-5, junk-7)
The Result received is:
['piece junk']

The second example:
The phone has bad screen. I find it disinteresting

The two rules are applied (Rule 2 and Rule 5).
The result received is:
['Screen bad,' 'Find disinteresting']

4.4 Summary

This research explored a rich set of syntactic rules and its associations. It has been proven that the usefulness of these syntactic rules and relations in the representing of the relationships between the product aspects and the corresponding opinions. In this research work, rule-based methodology covering syntactical dependency relations will help in guiding the extraction process of thematic-opinion word pairs.

Better accuracy was achieved with the proposed techniques, which is shown in Chap. 7 by exploring different datasets than the existing dependency models to extract aspect-based opinions from reviews (customers).

The experiments showed that proposed technique achieves better accuracy than existing dependency models for aspect-based opinion mining from customer reviews. Lastly, this approach can be improved by applying more rules for aspect–opinion relation extraction. As a possible direction for future work, finding more useful dependencies along with expanding the opinion lexicon might be considered.

References

1. Rana, T. A., & Cheah, Y. N. (2016). Aspect extraction in sentiment analysis: comparative analysis and survey. *Artificial Intelligence Review* 1–25.
2. Mukherjee, S., & Joshi, S. (2013). Sentiment aggregation using ConceptNet ontology. In *IJCNLP* (pp. 570–578).
3. Liu, H., & Singh, P. (2004). ConceptNet—A practical commonsense reasoning tool-kit. *BT Technology Journal, 22*(4), 211–226.
4. Miller, G. A., Beckwith, R., Fellbaum, C., Gross, D., & Miller, K. J. (1990). Introduction to WordNet: An on-line lexical database. *International Journal of Lexicography, 3*(4), 235–244.
5. Blair-Goldensohn, S., Hannan, K., McDonald, R., Neylon, T., Reis, G. A., & Reynar, J. (2008). Building a sentiment summarizer for local service reviews. In *WWW workshop on NLP in the information explosion era* (Vol. 14, pp. 339–348).
6. Zhang, W., Yu, C. & Meng, W. (2007). Opinion retrieval from blogs. In *Proceedings of the sixteenth ACM conference on conference on information and knowledge management* (pp. 831–840), New York, NY, USA: ACM.
7. Lee, D., Jeong, O. & Lee, S. (2008). Opinion mining of customer feedback data on the web. In *Proceedings of the 2nd international conference on ubiquitous information management and communication* (pp. 230–235), New York, NY, USA: ACM.
8. Lazhar, F., & Yamina, T. G. (2016). Mining explicit and implicit opinions from reviews. *International Journal of Data mining, Modelling and Management, 8*(1), 75–92.
9. Manning, C. D., Raghavan, P., & Schütze, H. (2008). *Introduction to information retrieval* (Vol. 1, No. 1, p. 496). Cambridge: Cambridge university press.

10. Tsytsarau, M., & Palpanas, T. (2012). Survey on mining subjective data on the web. *Data Mining and Knowledge Discovery, 24*(3), 478–514.
11. Bhatia, S., Sharma, M., & Bhatia, K. K. (2016). A novel approach for crawling the opinions from World Wide Web. *International Journal of Information Retrieval Research (IJIRR), 6*(2), 1–23. (ESCI, WEB OF SCIENCE).
12. Bhatia, S., Sharma, M., & Bhatia, K. K. (2018). Sentiment analysis and mining of opinions. In *Internet of things and big data analytics toward next-generation intelligence* (pp. 503–523). Cham: Springer.
13. Chen, D., & Manning, C. D. (2014). A fast and accurate dependency parser using neural networks. In *EMNLP* (pp. 740–750).
14. Dey, N., Wagh, S., Mahalle, P. N., & Pathan, M. S. (2019). *Applied machine learning for smart data analysis.* CRC Press.
15. Bhatia, S., Sharma, M., Bhatia, K. K., & Das, P. (2018). Opinion target extraction with sentiment analysis. *International Journal of Computing, 17*(3), 136–142. (Elsevier, SCOPUS).
16. Marrese-Taylor, E., Velásquez, J. D., Bravo-Marquez, F., & Matsuo, Y. (2013). Identifying customer preferences about tourism products using an aspect-based opinion mining approach. *Procedia Computer Science, 22*, 182–191.
17. Popescu, A. M., Nguyen, B., & Etzioni, O. (2005). OPINE. In *Proceedings of HLT/EMNLP on interactive demonstrations, association for computational linguistics* (pp. 32–33).
18. Hu, M., & Liu, B. (2004a). Mining opinion features in customer reviews. In *AAAI* (Vol. 4, No. 4, pp. 755–760).
19. Ali, M. N. Y., Sarowar, M. G., Rahman, M. L., Chaki, J., Dey, N., & Tavares, J. M. R. (2019). Adam deep learning with SOM for human sentiment classification. *International Journal of Ambient Computing and Intelligence (IJACI), 10*(3), 92–116.

Chapter 5
Opinion Classification

Classifying opinions have been identified as prime task of sentiment. In many applications of NLP, classifying text has been considered an essential component. Traditionally, text classification was mainly done by manual engineering tasks such as developing handcrafted features by consulting dictionaries, knowledge-based techniques or customized hierarchical components (tree kernels). It is mainly achieved by humans. These methods have not proved to be much efficient in today's fast - developing phase as manual feature engineering is required. The work discusses in detail the classification techniques; models an also explain the concepts of deep learning for classifying text that follows the least intervention of humans. This chapter aims at presenting classified opinions to the user's query by defining aspect-based binary classification problem. A novel technique is proposed which transforms input text into binary form of vectors followed with the application of CNN. In this chapter, CNN is used in which classifiers are automatically trained and learning is achieved through word representations. The novel way to develop semantically rich input vector is discussed with the application of CNN. The classification model is developed, and the performance is evaluated with different architectural decisions and with varying choice of hyperparameter settings which is discussed in Chap. 7.

5.1 Sentiment Analysis and Opinion Classification

Opinions are used to express reviews that provide information about how a product is perceived. People's contributions lie in posting text messages in the form of opinions and emotions on different topics such as movie, book, product, and politics. The reviews available online can be available in thousands, so making the right decision to select a product becomes a very tedious task, and it can be achieved by viewing the classified results of opinions as positive or negative. The common approach to comprehend opinion is viewing it with both perspectives, i.e., positive and negative,

© The Author(s), under exclusive license to Springer Nature Singapore Pte Ltd. 2020
S. Bhatia et al., *Opinion Mining in Information Retrieval*,
SpringerBriefs in Computational Intelligence,
https://doi.org/10.1007/978-981-15-5043-0_5

and the process is called as 'sentiment analysis.' It has been gaining importance over last few years from academics and industry. The methods that exploit prior knowledge from large sets of unlabelled texts are helpful in achieving scalability and also crossing the bridge that exists between contextual information and sentiment analysis. The traditional methods discussed previously in sentiment analysis are confined to manual feature engineering tasks, and different machine learning classifiers like NB, SVM, DT, ME are applied on these features [1] for the classification task. Salient features can be captured for this difficult task through deep learning which has proved to be successful in classification, and CNN has given remarkable results [2, 3]. In recent years, deep learning models have given successful results in image processing, computer vision [4], and speech recognition [5]. It is growing tremendously in NLP in text classification as well. Different deep learning methods have been suggested to study word vector representations using NN [6–8]. Massive amount of data needs training, and a large numbers of parameters need to be considered for building the network. The proposed model has taken inspiration from image classification technique proposed by Razavian et al. [9]. The deep learning model gives outstanding performance on extracting features on a variety of tasks. Schmidhuber in 2015 had reviewed concepts and theories behind deep learning in NN and gave a detailed idea of how deep learning can be used [10].

5.1.1 Problem in AI Context

The field of sentiment analysis, which also known as opinion mining, analyzes the sentiments, attitudes, evaluations, emotions, appraisals, opinions, and reviews towards entities such as services, issues, topics, organizations, individuals, events, products, and their attributes. So it is a large problem domain to explore. The umbrella of sentiment analysis or opinion mining comprises different tasks like the opinion mining, opinion extraction, sentiment analysis, sentiment mining, subjectivity analysis, effect analysis, review mining emotion analysis. Sentiment analysis is basically a natural language processing (NLP) problem and touches every not solved problems of NLP like negation handling, word sense disambiguation, and conference resolution. However, 'sentiment analysis is a highly restricted NLP problem' this could be a useful fact as semantics of each document or sentence need not to understand fully, only few aspects required. So the sentiment analysis enables us to explore rich set of interrelated subproblems and allow us to abstract a structure from the complex and intimidating unstructured natural language text. This enables the researchers to design more accurate and robust system by exploiting the correlation of these subproblems from practical point of view. Opinions and sentiments are subjective which allow examining a collection of opinions of many people rather than single person's view. These views can be acquired from different resources like news articles, tweets (Twitter postings), forum discussions, blogs, and Facebook postings. This makes the large collection of opinions and sentiments on the Web which incline the research towards the summarization of these opinions [11]. This summarization is application-oriented like what is the subjectivity and emotions, what are the key concepts, etc. Sentiment analysis can be modeled as machine learning problem of classification, and two subproblems must be resolved (1) subjectivity classification problem, and (2) polarity classification problem. The former one takes the sentences as subjective or objective of classification and the later one classifies the sentence as positive, negative or neutral is known as polarity. The sentiment analysis is a multi-faceted problem with many challenging subproblems, and this chapter explores the machine learning as well as deep learning solution techniques for the same.

5.1.2 Document-Level Classification

The polarity classification of a document is also known as document-level classification as it aims to classify whole document as positive, negative, or neutral. Formally, the given document D for the evaluation of entity E can be classified according to overall sentiment S of the opinion holder H about the entity E. The classification is irrelevant of the entity E, time of opinion t, and opinion holder H. The value of S can be positive or negative which makes sentiment analysis as classification problem. Same becomes regression problem when S takes ordinal scores like 1 to 10. Assigning one sentiment orientation to entire document is not a practical approach, as document evaluates more than one entity at the same time. The supervised learning, unsupervised learning, and deep learning techniques are implemented for document-level classification. The supervised classification of document-level classification is very popular and most common technique. Recently, several extensions to this research have also appeared, most notably, cross-domain sentiment classification (or domain adaptation) and cross-language sentiment classification, which will also be discussed at length.

5.1.3 Sentence Subjectivity

The subjectivity and emotion are two important perceptions closely related to opinion and sentiments. Subjectivity of a sentence expresses the personal feelings, beliefs, and views of user about the world. This differs from the objective of the sentence which is factual information. For example, 'iMac is an Apple company product' is an objective sentence and 'I prefer iMac' is a subjective sentence. The subjectivity can be expressed in many forms, e.g., allegations, opinions, beliefs, speculations, desires, and suspicions [12, 13]. Subjectivity can be sometimes confused with opinionated. When a document implies or expresses a negative or positive sentiment that means it is opinionated. Subjectivity classification is classifying the sentence into subjective or objective sentence. No sentiment is expressed in a subjective sentence, e.g., 'I think that she went home.' However, objective sentences include opinions or sentiments due to desirable or undesirable facts, e.g., 'The phone speaker was not working' [14].

The implications or opinions can be implicit or explicit in the sentence and could be positive or negative. Explicit opinion bearing subjectivity in sentence is a type of subjectivity. Other than this judgment, affect, appreciation, hedge, speculation, perspective, agreement and disagreement, political stances and arguing are other type of subjectivity which not has been studied extensively. They may or may not imply the sentiments [11–14]. Emotions, another topic will be discussed in detail following sections. Emotions and opinions are not equivalent terms but interrelated in many aspects.

5.2   Opinion Strength and Polarity Generation

Sentiment classification is concerned with finding the contextual polarity of a sentence. A majority of research work on opinion mining is considered word as an essential sentiment bearing element [15–19]. Researchers have done two primary tasks at word-level opinion mining [20], i.e., subjectivity analysis and opinion polarity determination. Subjectivity analysis is concerned with examining the subjectivity of the word, i.e., whether the word is opinionated or objective. The opinion polarity determination is concerned with finding the semantic orientation of such subjective word, i.e., whether it communicates a positive opinion or negative opinion. Most of the research concentrates on finding semantic orientation of each word of the review. Sentence- or document-level opinion mining is accomplished by merging of word- or phrase-level opinion information. The authors [21] stated that adjective words present in document are strong opinion bearing words. The authors contended that descriptive words are solid indicators of estimations. Kamps et al. [22] built a graph between two seed words *good* and *bad* and target adjective words by utilizing WordNet synonymy associations. Many researchers ignored other parts of speech like noun and adverb, which can also act as opinion bearing words. Riloff et al. [23], Kim and Hovi [24] considered parts of speech like noun and adverb words as opinion bearing words in their research work. The paper [23] concentrated on retrieving subjective nouns from documents. The existing methodologies of sentiment classification are arranged into lexicon and machine learning-based approaches. Different techniques are shown in Fig. 5.1 [25].

5.2.1   Dictionary-Based Approaches

Measuring the sentiment of public in real time has always been a difficult task. Previously people had to conduct a large amount of survey from significant number of people to know about the quality of a product. Traditional machine learning classifiers have shown good performance accuracy over lexicon-based approaches in text classification [26–30]. The machine learning techniques and classifiers like NB, ME, SVM have been used in both determining sentiments and categorizing texts by

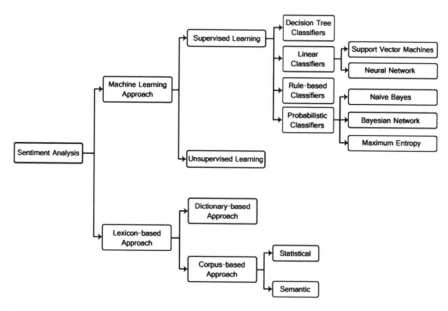

Fig. 5.1 Sentiment classification techniques

many researchers. The accuracy will be affected by choosing relevant features and applying the appropriate pre-processing tasks, which makes the system liable.

Hence, problem of pre-processing and feature engineering specific to the text genre exists [31]. All the above limitations widen up the scope of improvement, which cannot be overlooked.

5.2.2   Machine Learning Techniques

Classification involves segregating opinions into classes. Classifying text deals with inputs as '*i*' and outputs as '*o*', where '*i*' is the opinion in the form of text and '*o*' is the vector or number [32]. The number of different classes '*C*' is mentioned according to the approximation function applied on the dataset. It can be stated as:

$$C(n) : i(n) \rightarrow o(n) \tag{5.1}$$

The *n*th instance of the training data is classified according to the problem, [*i*(*n*), *o*(*n*)]. For example, 'The iPhone has poor battery backup' is the input which needs to be divided into either in positive (1) or negative (0) classes. The output will be 0 in case of the above-mentioned example.

The classification can be listed in the following three categories:

Binary classification: It classifies as positive class or negative class, $C = [Cpos, Cneg]$. The classifier decides which of the two available classes should be assigned to an instance of data. In this specific case, the classes are often referred to as a negative and a positive class.

Multi-class Classification: The classifier chooses among many available classes but only one class can be assigned at one time.

$$C = [C_1, C_2, \ldots C_n] \qquad (5.2)$$

Multi-label Classification: The classifier chooses among many available classes but simultaneously many classes can be assigned to one instance at a time.

In broader terms, sentiment classification techniques are mainly divided into machine learning approach and lexicon-based approach. The hybrid approach can form the combination of the above-listed two approaches [33, 34].

Supervised Learning: This method is based on the concept of learning from previous supervised samples. In other words, it learns by providing training. There are two sets of data: one is the training set and other is the test set. In supervised learning, the idea is that the highest possible accuracy is achieved by learning from a set of labeled examples provided in the training set to determine the unlabelled examples in the test set [35]. The objective of the learner (typically, a computer program) is to develop a method or a rule so that it can classify the examples present in the test set by considering the examples with class labels. The example of input–output pairs is given which is a function learnt by the algorithm. Definition: 'n' input–output pairs are provided as training examples such as $\{(x_1, y_1), (x_2, y_2) \ldots (x_n, y_n)\}$, learning algorithm works by considering a function f as:

$$f : X \rightarrow Y \qquad (5.3)$$

where X is input space and Y is output space. f is referred to as hypothesis, and function is used to map input to output space.

Unsupervised Learning: Unsupervised learning makes use of available lexical resources. The example of clustering comes as an application of the above technique [36].

Reinforcement Learning: Reinforcements are considered as a series of feedbacks on which algorithm learns and rewards or punishments are given as the model learns from its decisions.

Supervised learning problem can be taken as classification problem, where algorithm classifies a set of data by providing instances. Traditional methods have limitations with unsupervised learning as the feature extraction is a handcrafted task and is totally dependent on the expert doing the job. Also, in the earlier works done by various researchers [36–39], it has been shown that supervised learning algorithms have given much better accuracy over unsupervised learning algorithms in text classification. There are various subproblems which can be discriminated.

5.2.3   Traditional Supervised Learning Models

Some of the models have been discussed here:

Naïve Bayes (NB): This classifier is the simplest and the most widely used probabilistic classification algorithm [40–42].

Maximum Entropy (ME): This is another model, which performs probabilistic classification, making use of the exponential model [43]. This technique has been proven to be effective in many NLP classification tasks [44] including sentiment analysis. It is applicable to real-life scenarios as there is no conditional independence assumption.

Support Vector Machines (SVM): Cortes and Vapnik in 1995 [45] have proved SVM to be highly effective for the categorization of documents [46] based on similar topics. As opposed to the probabilistic classifiers like the previous two [47, 48], this method aims to find large margin between the different classes.

Neural Networks (NN): The tools designed to achieve computational tasks, modeled after brains are artificial neural networks (ANN). NN has shown good applicability in the areas of image processing and pattern recognition and is now becoming popular for solving NLP problems [49, 50]. The most common structure of NN is shown in Fig. 5.2. The action of this NN is determined by the weights applied in the hidden and output nodes [51].

NN has evolved as a promising substitute for different traditional classification methods. NN is susceptible for noisy data, gives excellent accuracy scores, and has become an important tool for classification.

5.2.4   What's Ahead

Deep learning has an edge over the traditional machine learning algorithms for sentiment classification because of its potential to overcome the challenges faced by it and handle the diversities involved, with good accuracy values and without the expensive demand for deciding over features that are important for classifying text. So, deep learning has emerged as a powerful tool for NLP in recent years. Some of the major advantages of deep learning are mentioned as:

- It does not require manual feature engineering.
- It allows good representation learning.
- It can handle the diverse tasks by making small changes in the system.

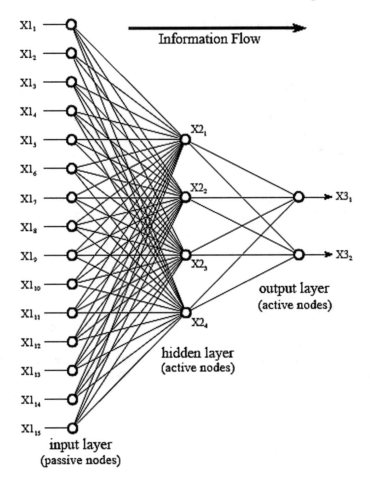

Fig. 5.2 Structure of neural network (NN)

5.3 Deep Learning in Opinion Mining

Deep learning is a powerful set of techniques for learning in NN [52]. Deep learning has recently shown much promise for NLP applications. One of the majorly evolving NLP applications is text classification, and it has given remarkable results in deep learning with multiple datasets [53, 54]. Other applications of deep learning with regard to NLP are language modeling (speech recognition, machine translation), acoustic modeling, chunking, named entity recognition, semantic role labeling, parsing, paraphrasing, question answering, and word-sense disambiguation. The motivational factors behind sentiment classification using deep learning models over other learning techniques are stated as under:

- Traditional classification requires the features to be selected/extracted from the raw data and that depends on the type of data used for building the model. Deep learning attempts to learn multiple levels of representations from the raw data by processing the data while going through many numbers of hidden layers.
- Another important difference between the two models is the number of layers of computations. The amount of data and number of layers required are more for deep learning architectures as compared to traditional machine learning architectures.
- Another important motivation for deep learning is that these architectures work extremely well with the unsupervised learning, i.e., (unlabeled examples) as well as supervised learning (labeled examples), while traditional methods have limitations with the unsupervised learning as the feature extraction is a handcrafted task totally dependent on the expert doing the job.
- Deep learning addresses the problem of vanishing error by adding a pre-training layer that is also referred to as the greedy layer. This layer is the unsupervised learning layer that learns before being moved up to the higher layers, then fine-tuning the weights from the previous unsupervised learning by using labeled data. In this way, it offers more abstract and hierarchical features moving through multiple hidden layers.
- Another feature that makes the deep learning different from the traditional machine learning methods is its ability to memorize and relate previous knowledge to the future information. The traditional methods are simply able to memorize the information up to certain layers without being able to correlate it with the past information.
- In NN, neurons are not sparse, so all the neurons actively participate; but in deep NN, some of the neurons are dropped out, because of the dropout feature. This helps in reducing the error more effectively.
- NN recursively diminishes the error through iterations by using some optimization techniques like gradient descent. Likewise in deep NN, the activation function will optimize and reduce the error over the iterations evolving them. Here the number of errors can be defined prior applying the activation function like, 50, 100, and so on.

There are various deep learning models which have shown good accuracy in text classification such as CNN, LSTM, and RNN.

In this chapter, CNN for opinion classification is used as it has the major advantage to focus on local area of the sentence and only one local feature is identified which contributes most to the particular sentence. Different number of filters and different number of filter sizes can be used to achieve optimization. The entire local features slide over the window and after combining all the features, activation function is applied, and the best local feature is identified.

Word Embeddings
As mentioned earlier, deep learning models work by taking input as word embedding instead of features and learning of the middle layers in deep neural network is

achieved during training. Therefore, the complete understanding of how words are changed into vectors is explained below.

Constructing the vectors in numerical form in a way that enables one to use them for vector-based machine learning algorithms or in other words to capture the characteristics and semantics of the words can be considered as word embedding [2]. Several approaches are widely adopted for word embedding. Some of them are explained below:

Bag of Words (BOW)

The text contains words (broken into tokens). These tokens are considered as n-grams which are used to encode a vector, where n denotes the number of tokens in text. The tokens are overlapped. For example, two sentences as shown below:

Sentence 1: I love to watch action movies.
Sentence 2: The hero of the picture was so amazing.

Since not all words that appear in the document are significant, so weighting features can be collaborated with bag of words by counting the number of word co-occurrences in between. BOW example is shown in Fig. 5.3.

$tf_{t,d}$ is known as *term frequency* which is used to compute the score by assigning weight to each term in the document. Weight depends on the number of times that the term has occurred in the document. The *term* is denoted by t, and *document* is denoted as d.

Using document-level statistic, df_t is known as *document frequency* and is defined as the number of documents in the collection that contain a *term* t [55].

Taking N as the total number of documents in a collection, the *inverse document frequency* (idf) of a *term* t is defined as follows:

Sentence 1

I love to watch action movies.

Sentence 2

The hero of the picture was very amazing.

Term	Sentence 1	Sentence 2
Love	1	0
Watch	1	0
Action	1	0
Hero	0	1
Picture	0	1
Movies	1	0
Very	0	1
Amazing	0	1

Stop word list

I
To
The
Of
was

Fig. 5.3 Example of bag of words (BOW)

$$idf_t = \log \frac{N}{df_t}.$$ (5.4)

The definitions of term *frequency* and inverse document. It is calculated by the formula:

$$Tf - idf_t = tf_{td} * \log N/df_t$$ (5.5)

Dimensionality reduction is important and can be achieved by involving matrices and using singular value decomposition (SVD), LSA, or deep learning word embedding [56].

Word Vectors
The word is kept in a higher dimensional vector space taking into account the words placed in the neighborhood. The drawback of the bag of words model is the sparsity which is unable to capture semantic information, so based on the cosine formula; distance between these vectors is calculated. Word2Vec [57] is a family of NN language models that learn a continuous vector representation of words—so-called word embeddings—given a context of some words with a shallow NN. Doc2Vec is a directly learned document vector which can be calculated on the basis of word vectors [58]. Networks are aware of only numbers and functions, so NN in NLP cannot accept raw words as input. Some transformation is necessary where words are can be changed to feature vectors called as word embedding [59]. The structure and distinct features of the words can be understood well with these vectors if extracted properly. Learning can also be achieved. Most of the researchers use pre-trained vectors released by Google which is about 100 billion words trained on part of Google News dataset. There are near 3 million words and phrases with 300-dimensional vectors distributed in space. NN takes these word vectors as inputs for NLP tasks. Cosine distance calculation is one of the factors for determining closeness of word vectors [59]. For example, embedding of words *teacher* and *student* is closer to that of *education* as compared to that of *entertainment*.

$$< king > + < queen > = < boy > + < girl >$$
$$< playing > + < played > = < trying > + < tried >$$
$$< Japan > + < Tokyo > = < Berlin > + < Germany >$$

The example shown above describes the utility of these vectors in NLP prediction tasks named as entity recognition and sentiment classification [60].

Convolutional Neural Networks
One of the algorithms which have shown good accuracy in text classification in deep learning is CNN. A unified NN architecture which was proposed by Collobert et al. in 2011 had applied to various NLP tasks [60]. The architecture, known as CNN, takes concatenated word vectors of the text as input and involves convolutional and max pooling layers prior to the general NN framework. The 'sentence vector' is then fed

into a fully connected NN with hidden layers and activation functions like softmax or sigmoid to ultimately reach the final layer whose size is the same as the number of labels.

Generally, for NLP tasks, window approach is used [60]. In this method, window size is made static and words are fed into the network for classification and to find the new neighboring words. [61] explored the performance of deep learning models and implemented NB classifier as a baseline for comparison. Another paper [62] discussed a wide range of NLP classification models and implemented supervised classifiers.

5.4 Aspect-Based Opinion Classification

Convolutional networks can be called as variations which are biologically inspired by multilayer perceptron (MLPs) as it consists of many identical copies of same neuron. CNN architecture is built upon the three main layers discussed below:

Convolutional Layers

Convolutions in the context of machine learning are filters that are applied to receptive fields of an input tensor [63], yielding tensors of filter maps representing the activation of a region on the inputs. In this case, the input tensors are simply matrices, and the individual feature maps are scalar due to the dot product of the *weight vector* w and the *sentence matrix* A. In classical applications of ConvNets on images, the input tensor has two dimensions for the pixels (width * height) and an additional dimension of Red Green Blue (RGB) values where each of the RGB dimensions is also known as a channel. The convolution is applied to a matrix by computing the filter on windows where each window represents a slice of the matrix as given by a stride width and height. Words in each row are reduced row-wise using stride, thus viewing words in atomic form. A filter of height h is applied to a window of words (i representing number of rows and j as number of columns).

Rows of the matrix represent a word in NLP applications.

$$i : j + h - 1 \tag{5.6}$$

Choosing a filter that is applied with a column-wise stride too can be used for convolutions on characters in the word [31]. A convolution is defined by a kernel function (activation function) that is applied to the patches as a weight matrix and a chosen nonlinearity.

$$f : c_j = f(w.A_{[i:j+h-1]} + b) \tag{5.7}$$

where f is a nonlinear function applied to the dot product of the filter and the word vector window. For each filter, the corresponding feature map of all regions is given as:

$$[c_1 \ldots \ldots c_{n-h+1}] = c \in R^{n-h+1} \qquad (5.8)$$

Each filter learns a feature map given by $c \in R^{n-h+1}$.

Pooling Layers
In computer vision tasks, pooling layers are introduced for invariance of the nodes to rotation, translation or scaling, modeling that an object is the same, even occurring on another patch of the original image, or in another size. Another handy property is that a variable-sized feature map is thus reduced to a fixed size entity. Let c_j be the jth feature map. The pooling function used by Kim [64] is defined as:

$$c_j = \max(c_j) \qquad (5.9)$$

where c_j can be interpreted as the most prominent feature of the feature map (shown in Fig. 5.4). Another handy attribute of pooling layers is that the network now can handle differently sized texts because despite the size of the input tensor or the then computed feature map the pooling operation yields only one value whatsoever [65]. Also, pooling layers can make features invariant to translation or scaling because of its independency of the feature's position in the feature map.

In the example given above, the colored window shows that the maximum among all the values is taken.

Dropout Dropout is a neat method to prevent overfitting in neural networks [66].

Fully Connected Layer The softmax classifier is used for binary classification.

The proposed technique is extended to include the model proposed by [64]. The classification model discussed by [64] is taken as a baseline model in many previous researches also [67, 68]. The work proposes to employ wide convolution instead of simple convolution. The proposed technique uses vectors which are made semantically rich by concatenating vectors obtained from SentiWordNet and Word2Vec embedding. The classification accuracy beats Kim's method due to the introduction

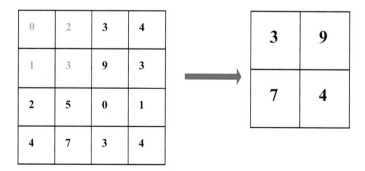

Fig. 5.4 Max pooling example (2 * 2 patches with a stride size of 2)

of wide convolution and a new univariate vector to the network. The flow of the model is given in Fig. 5.5.

The proposed CNN follows the pattern in Table 5.1.

The proposed work includes feature-level sentiment classification and document-level sentiment classification. For document-level sentiment classification, the user will receive the positive reviews and negative reviews against the particular dataset. On the other hand, in feature-level sentiment classification, the positive and negative reviews will be received on the particular feature by specifying the aspect explicitly. The ConvNet works by taking input in the form of vectors. The size of the input in CNN is calculated according to the dataset. Generally, the average sentence length is taken to be of 11 words. Two techniques are employed for getting a univariate vector; one vector is taken from SentiWordNet score, and the second vector is built from

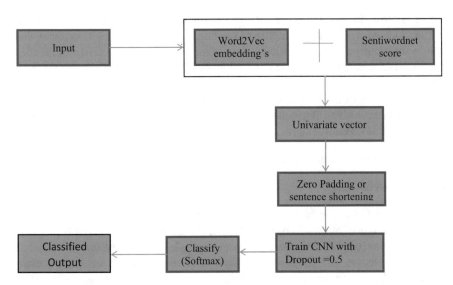

Fig. 5.5 Basic flow of CNN for classifying opinions

Table 5.1 Proposed CNN Pattern

Input
$\rightarrow$ [(Convolutional $\rightarrow$ Activation) $\rightarrow$ Pooling]
(FC $\longrightarrow$ Dropout) $\rightarrow$ Activation)
Output

the text using Word2Vec embeddings. In many previous CNN models discussed [2, 3, 64], the input is generally in the form of Word2Vec. The proposed CNN model discusses a novel technique for building the univariate vector. Here, instead of Word2Vec input, another vector containing the SentiWordNet score is appended to the input vector and sent as an input to the model. Padding is done to achieve matrix compliance.1 Max pooling is applied to sample down different length vectors into the same length before fully connected layer. Hence, it is applied after top convolution layer. The probability distribution of the input data is also tested with LR and SVM. The adadelta update rule [69] is used to determine the learning rate.

5.5 Summary

This chapter proposed the novel algorithm using grammar dependency rules for obtaining aspects from opinions. The proposed technique has selected a good set of the given rules, and conducting various experiments on different datasets have proved that our approach has performed much better extraction than the previous papers. The experimental results demonstrated its superior performance and its effectiveness. Opinion classification is considered a second problem which is achieved by CNN. The work in the chapter proposes to add a semantic layer above the CNN layer which helps in enriching the word vector semantically. The proposed CNN model is able to learn phrase-level features automatically which can be used to achieve classification tasks efficiently. The comparison of the traditional classifiers with the softmax classifier in CNN model also gives a clear idea that goes in line with the deep learning revolution. The accuracy results in this research outperform the previously built CNN models in text classification. Therefore, it is highly expected that training with a deeper CNN and a bigger dataset will produce better results.

References

1. Liu, B. (2015). *Sentiment analysis: Mining opinions, sentiments, and emotions*. Cambridge University Press.
2. Mikolov, T., Sutskever, I., Chen, K., Corrado, G. S., & Dean, J. (2013a). Distributed representations of words and phrases and their compositionality. In *Advances in Neural Information Processing Systems* (pp. 3111–3119).
3. Severyn, A., & Moschitti, A. (2015). UNITN: Training Deep Convolutional Neural Network for Twitter Sentiment Classification. In *SemEval@ NAACL-HLT* (pp. 464–469).
4. Krizhevsky, A., Sutskever, I., & Hinton, G. (2012). ImageNet classification with deep convolutional neural networks. In *Proceedings of NIPS 2012*.
5. Graves, A., Mohamed, A., & Hinton, G. (2013). Speech recognition with deep recurrent neural networks. In *Proceedings of ICASSP 2013*.
6. Bengio, Y., Ducharme, R., & Vincent, P. (2003). Neural probabilitistic language model. *Journal of Machine Learning Research, 3,* 1137–1155.

7. Yih, W., Toutanova, K., Platt, J., & Meek, C. (2011). Learning discriminative projections for text similarity measures. In *Proceedings of the Fifteenth Conference on Computational Natural Language Learning* (pp. 247–256).

8. Mikolov, T., Sutskever, I., Chen, K., Corrado, G., & Dean, J. (2013). Distributed representations of words and phrases and their compositionality. In *Proceedings of NIPS 2013*.

9. Sharif Razavian, A., Azizpour, H., Sullivan, J., & Carlsson, S. (2014). CNN features off-the-shelf: an astounding baseline for recognition. In *Proceedings of the IEEE Conference on Computer Vision and Pattern Recognition Workshops* (pp. 806–813).

10. Schmidhuber, J. (2015). Deep learning in neural networks: An overview. *Neural Networks, 61,* 85–117.

11. Hu, M., & Liu, B. (2004). Mining and summarizing customer reviews. In: *Proceedings of ACM SIGKDD International Conference on Knowledge Discovery and Data Mining (KDD-2004).*

12. Riloff, E., Patwardhan, S., & Wiebe, J. (2006). Feature subsumption for opinion analysis. In: *Proceedings of the Conference on Empirical Methods in Natural Language Processing EMNLP-2006.*

13. Wiebe, J. (2000). Learning subjective adjectives from corpora. In *Proceedings of National Conference on Artificial Intelligence (AAAI-2000).*

14. Zhang, L., & Liu, B. (2011). Identifying noun product features that imply opinions. In *Proceedings of the Annual Meeting of the Association for Computational Linguistics (short paper) (ACL-2011).*

15. Pang, B., & Lee, L. (2008). Opinion mining and sentiment analysis. *Foundations and Trends® in Information Retrieval, 2*(1–2), 1–135.

16. Liu, B. (2007). *Web data mining: Exploring hyperlinks, contents, and usage data.* Springer Science & Business Media.

17. Wei, W. (2011). Analyzing text data for opinion mining. *Natural Language Processing and Information Systems*, (pp. 330–335).

18. Zhu, J., Wang, H., Zhu, M., Tsou, B. K., & Ma, M. (2011). Aspect-based opinion polling from customer reviews. *IEEE Transactions on Affective Computing, 2*(1), 37–49.

19. Prabowo, R., & Thelwall, M. (2009). Sentiment analysis: A combined approach. *Journal of Informetrics, 3*(2), 143–157.

20. Nasraoui, O. (2008). Web data mining: Exploring hyperlinks, contents, and usage data. *ACM SIGKDD Explorations Newsletter, 10*(2), 23–25.

21. Hatzivassiloglou, V., & McKeown, K. R. (1997). Predicting the semantic orientation of adjectives. In *Proceedings of the Eighth Conference on European Chapter of the Association for Computational Linguistics, Association for Computational Linguistics* (pp. 174–181).

22. Kamps, J., Marx, M., Mokken, R. J., & De Rijke, M. (2004). Using WordNet to Measure Semantic Orientations of Adjectives. In *LREC,* (Vol. 4, pp. 1115–1118).

23. Riloff, E., Wiebe, J., & Wilson, T. (2003). Learning subjective nouns using extraction pattern bootstrapping. In *Proceedings of the Seventh Conference on Natural Language Learning at HLT-NAACL 2003, Association for Computational Linguistics* (Vol. 4, pp. 25–32).

24. Kim, S. M., & Hovy, E. (2004). Determining the sentiment of opinions. In *Proceedings of the 20th International Conference on Computational Linguistics* (pp. 1367). Association for Computational Linguistics.

25. Medhat, W., Hassan, A., & Korashy, H. (2014). Sentiment analysis algorithms and applications: A survey. *Ain Shams Engineering Journal, 5*(4), 1093–1113.

26. Pang, B., Lee, L., & Vaithyanathan, S. (2002). Thumbs up? Sentiment classification using machine learning techniques. In *Proceedings of the ACL-02 Conference on Empirical Methods in Natural Language Processing* (Vol. 10, pp. 79–86). Association for Computational Linguistics.

27. Xia, R., Zong, C., & Li, S. (2011). Ensemble of feature sets and classification algorithms for sentiment classification. Information Sciences, 181(6), (pp. 1138–1152).

28. Melville, P., Gryc, W., & Lawrence, R. D. (2009). Sentiment analysis of blogs by combining lexical knowledge with text classification. In *Proceedings of the 15th ACM SIGKDD International Conference on Knowledge Discovery and Data Mining* (pp. 1275–1284). ACM.

29. Zhang, Z., Ye, Q., Zhang, Z., & Li, Y. (2011). Sentiment classification of Internet restaurant reviews written in Cantonese. *Expert Systems with Applications, 38*(6), 7674–7682.
30. Ye, Q., Zhang, Z., & Law, R. (2009). Sentiment classification of online reviews to travel destinations by supervised machine learning approaches. *Expert Systems with Applications, 36*(3), 6527–6535.
31. Singhal, P., & Bhattacharyya, P. (2016). Sentiment analysis and deep learning: A survey.
32. Flach, P. (2012). *Machine learning: The art and science of algorithms that make sense of data.* Cambridge University Press.
33. Maynard, D., & Funk, A. (2011). Automatic detection of political opinions in tweets. In *Extended Semantic Web Conference* (pp. 88–99). Berlin, Heidelberg: Springer.
34. Russell, S., Norvig, P., & Intelligence, A. (1995). *A modern approach. Artificial Intelligence.* Egnlewood Cliffs: Prentice-Hall, 25, 27.
35. Learned-Miller, E. G. (2014). *Introduction to Supervised Learning* (Doctoral dissertation, PhD thesis), University of Massachusetts, Amherst, 2014. http://people.cs.umass.edu/~elm/Teaching/Docs/supervised2014a.pdf.
36. Feldman, R. (2013). Techniques and applications for sentiment analysis. *Communications of the ACM, 56*(4), 82–89.
37. Cambria, E., Schuller, B., Xia, Y., & Havasi, C. (2013). New avenues in opinion mining and sentiment analysis. *IEEE Intelligent Systems, 28*(2), 15–21.
38. Tsytsarau, M., & Palpanas, T. (2012). Survey on mining subjective data on the web. *Data Mining and Knowledge Discovery, 24*(3), 478–514.
39. Zhai, Z., Liu, B., Wang, J., Xu, H., & Jia, P. (2012). Product feature grouping for opinion mining. *IEEE Intelligent Systems, 27*(4), 37–44.
40. Russell, S. J., & Stuart, J. (2003). Norvig. *Artificial Intelligence: A Modern Approach* (pp. 111–114).
41. McCallum, A., & Nigam, K. (1998). A comparison of event models for naive bayes text classification. In *AAAI-98 Workshop on Learning for Text Categorization* (Vol. 752, pp. 41–48).
42. Chen, S. F., & Rosenfeld, R. (2000). A survey of smoothing techniques for ME models. *IEEE transactions on Speech and Audio Processing, 8*(1), 37–50.
43. Jaynes, E. T. (1957). Information theory and statistical mechanics. *Physical Review, 106*(4), 620.
44. Berger, A. L., Pietra, V. J. D., & Pietra, S. A. D. (1996). A maximum entropy approach to natural language processing. *Computational linguistics, 22*(1), 39–71.
45. Vapnik, V., & Cortes, C. (1995). Support vector networks. *Machine Learning 20*, 273–297.
46. Joachims, T., Nedellec, C., & Rouveirol, C. (1998). Text categorization with support vector machines: Learning with many relevant. In *10th European Conference on Machine Learning.*
47. Rennie, J. D., Shih, L., Teevan, J., & Karger, D. R. (2003). Tackling the poor assumptions of naive bayes text classifiers. In *ICML* (Vol. 3, pp. 616–623).
48. Yang, N., Dey, N., Sherratt, S., & Shi, F. (2019). Emotional state recognition for AI smart home assistants using Mel-Frequency Cepstral coefficient features. *Journal of Intelligent & Fuzzy Systems.*
49. Bhatia, S., Sharma, M., & Bhatia, K. K. (2018). Sentiment analysis and mining of opinions. In *Internet of Things and Big Data Analytics Toward Next-Generation Intelligence* (pp. 503–523). Cham: Springer.
50. Dey, N., Ashour, A. S., Fong, S. J., & Bhatt, C. (eds.) (2018). *Healthcare data analytics and management.* Academic Press.
51. Smith, W. S. (1997). Neural Networks. In *Digital Signal Processing* (pp 451–480), San diego, CA: California Technical Publishing.
52. Deng, L., & Yu, D. (2014). Deep learning: Methods and applications. *Foundations and Trends in Signal Processing, 7*(3–4), 197–387.
53. Chaudhary, P., & Agrawal, R. (2019). A comparative study of linear and non-linear classifiers in sensory motor imagery based brain computer interface. *Journal of Computational and Theoretical Nanoscience, 16*(12), 5134–5139.

54. Chaudhary, P., & Agrawal, R. (2018). Emerging threats to security and privacy in brain computer interface. *International Journal of Advanced Studies of Scientific Research, 3*(12), 340–344 (2018). [Online]. Available: https://ssrn.com/abstract=3326692.
55. De Marneffe, M. C., & Manning, C. D. (2008). The Stanford typed dependencies representation. In: *Coling 2008: Proceedings of the Workshop on Cross-framework and Cross-domain Parser Evaluation* (pp. 1–8). Association for Computational Linguistics.
56. Bengio, Y., Ducharme, R., Vincent, P., & Jauvin, C. (2003). A neural probabilistic language model. *Journal of Machine Learning Research, 3*(Feb), 1137–1155.
57. Mikolov, T., Karafiát, M., Burget, L., Cernocký, J., & Khudanpur, S. (2010). Recurrent neural network based language model. In *Interspeech* (Vol. 2, p. 3).
58. Le, Q., & Mikolov, T. (2014). Distributed representations of sentences and documents. In *Proceedings of the 31st International Conference on Machine Learning (ICML-14)* (pp. 1188–1196).
59. Mikolov, T., Chen, K., Corrado, G., & Dean, J. (2013b). Efficient estimation of word representations in vector space. arXiv preprint arXiv:1301.3781.
60. Collobert, R., Weston, J., Bottou, L., Karlen, M., Kavukcuoglu, K., & Kuksa, P. (2011). Natural language processing (almost) from scratch. *Journal of Machine Learning Research, 12*(Aug), 2493–2537.
61. Shirani-Mehr, H. (2014). Applications of deep learning to sentiment analysis of movie reviews.
62. Pouransari, H., & Ghili, S. (2014). Deep learning for sentiment analysis of movie reviews.
63. Dullemond, K., & Peeters, K. (1991). *Introduction to tensor calculus.* KeesDullemond and Kasper Peeters.
64. Kim, Y. (2014). *Convolutional neural networks for sentence classification.* arXiv preprint arXiv:1408.5882.
65. Schulze, C., (2016). Machine Learning for Atmosphere Classification of Movies. (Thesis). Technische Hochschule Mittelhessen.
66. Srivastava, N., Hinton, G., Krizhevsky, A., Sutskever, I., & Salakhutdinov, R. (2014). Dropout: A simple way to prevent neural networks from overfitting. *The Journal of Machine Learning Research, 15*(1), 1929–1958.
67. Mandelbaum, A., & Shalev, A. (2016). Word embeddings and their use. In *Sentence Classification Tasks.* arXiv preprint arXiv:1610.08229.
68. Zhang, Y., & Wallace, B. (2015). A sensitivity analysis of (and practitioners' guide to) convolutional neural networks for sentence classification. arXiv preprint arXiv:1510.03820.
69. Zeiler, M. D. (2012). Adadelta: An adaptive learning rate method. arXiv preprint arXiv:1212.5701.

Chapter 6
Opinion Summarization

Opinion summarization is a prime and an important step in 'opinion mining.' Summarization helps in reducing the text in the shortest way possible such that important information can be gained from the text, without any loss of information and significant properties in the text remain preserved. With the huge amount of reviews posted online, a summary is thus required, to influence a person in making correct decision considering all its important aspects. The task of automatic summarization has increased interest among its communities of NLP and text mining.

In this chapter, both abstractive summarization and extractive summarization are accomplished using two different techniques, graph-based method and PCA method, respectively.

6.1 Text Summarization

Since the volume of content is increasing and the reviews, feedbacks, and opinions shared by different people are expressed again and again. Summarizing text is taken as an interesting task of NLP [1, 2]. The intersection of two fields, computer science and linguistics, leads to the formation of NLP. In other words, thoughts and notions can be swapped over between human and computers by applying NLP on processed data. Lexical analysis, syntax analysis, semantic analysis, and discourse processing are the main components of NLP. Multiple opinions are required to generate summary and it is based on feature selection [3], feature rating, [4], and identifying sentence that contains features [5]. The task of summarization is given in Fig. 6.1.

Primarily summarization is categorized into two types, but there may be other approaches such as graph-based, general, and hybrid techniques given in Fig. 6.2 [6].

© The Author(s), under exclusive license to Springer Nature Singapore Pte Ltd. 2020
S. Bhatia et al., *Opinion Mining in Information Retrieval*,
SpringerBriefs in Computational Intelligence,
https://doi.org/10.1007/978-981-15-5043-0_6

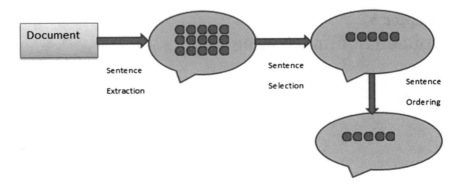

Fig. 6.1 Task of summarization

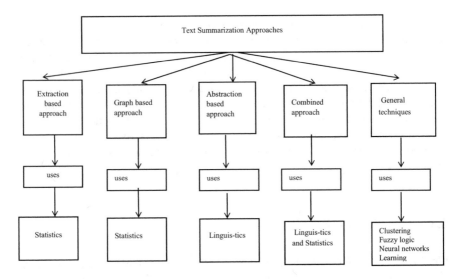

Fig. 6.2 Text summarization approaches

6.1.1 *Extractive Summarization*

Extractive summarization involves combining all the extracts taken from the corpus into summary. It helps the user by extracting the most important pieces of information from the huge corpus [7, 8]. It is verbose as non-essential parts of a sentence also get included only when data is redundant. It suffers from the dangling problem as it solely lies in the content and extracting sentences. So, sometimes irrelevant sentences get included and the important part of sentences such as pronouns is left out, those otherwise, needs to be conserved [9].

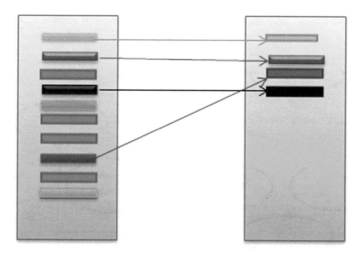

Fig. 6.3 Extractive summarization

Single-document summarization focuses on extractive techniques like machine learning methods—NB, DT, hidden Markov model, log-linear, and NN. Extractive summarization creates summary in copy paste fashion as shown in Fig. 6.3.

6.1.2 Abstractive Summarization

Abstractive summarization involves generating new cohesive text that may not be present in the original information. It requires deep learning over the text to determine the meaning of each word and phrase in order to generate summary [10]. It builds a semantic representation of the text from which summary gets generated. In Fig. 6.4, lines show how the merging is done to form a new cohesive text. In the semantic representation, graph nodes are concepts and edges are semantic relations. Example:

Sentence A: Mouth-watering food is served.
Sentence B: Staff serves good food in the hotel.
Summary Generated: Staff serves good mouth-watering food in the hotel.

Advanced language generation techniques are needed to produce a grammatically oriented summary highlighting the 'form.' Multi-document summarization follows abstractive approaches. Automatic summarization is involved either using statistical approach, linguistic approach, or using a hybrid approach. The sentence is ranked on the basis of keywords and determination of probability of key terms through weight by statistical approach. On the other hand, the linguistic approach examines text and finds out the concepts and associations between sentences by looking into semantics by using POS tagging, analyzing grammar, and dictionary meaning of

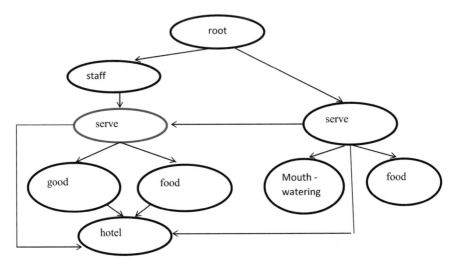

Fig. 6.4 Semantic representation of the text for abstractive summarization

relevant sentences [11]. Abstractive techniques make use of linguistic or hybrid approaches, but extractive technique makes use of statistical approach. Computation is stronger in extractive approaches but better results on summarization are produced by abstractive approach [12]. There are hybrid approaches also that tend to merge both statistical-based and linguistic-based methods. Large numbers of researchers have proposed summarization methods based on NLP. Because of the reasons listed above, the previous other researches also demonstrated that abstractive summaries can give much better results when one has to summarize product reviews, blogs, articles, etc., than extractive summaries [13].

6.2 Traditional Approaches

The extensive study has been done on abstractive-based technique rather than extractive technique. The rule-based approach [14], sentence compression [15–17], merging sentence based on their semantics [18, 19], etc., have been studied. The authors [20] proposed an effective ranking method for summarizing text on using hash algorithms for constructing graphs. The researchers [21] developed the novel method of generating abstractive summary using the directed graphs. The use of connectors by providing input in the form of graphical form helps in the reduction of redundant sentences as opinions [22, 23]. However, the proposed work produces readable, concise, and fairly well-formed summaries but still has a limitation mark. The unavailability of a preexisting connector might not be able to fuse the sentences which are capable of getting that stacking up together. The complexity of this approach is high as it lays too much emphases on the surface order of words. The papers [24] proposed the graphical-based summary by incorporating the rules of natural language processing. The methodology describes by building the word graph, ensuring the correctness by imposing POS constraints and the scoring of paths based on the calculation of overlap. Finally, the fusion of sentences is achieved and the results are analyzed by ROUGE tool. The work limits to combining of the sentences that are semantically related but not related syntactically. The authors of [6] focused on a method for automatic text summarization using soft computing approach. Title and semantic similarity are primarily used for generating summary. The author developed the NLP parser and used human summarization rules: subject, verb, and object (SVO) for achieving the task of summarizing text. The proposed algorithm is not able to cope up with complex sentences as the rules stated are not enough to support these errors. Also, the problem of tag-based ambiguity needs to be resolved. The researchers [25] discussed the summarization process of large corpora using PCA. The author well formulated the low-rank approximation of a Salton matrix by taking text documents and running the procedure over the collection of news articles. Although the author tried to explain well the benefits of sparse PCA over normal PCA, still the paper did not explain well mathematically how encoding of the Salton matrix is achieved and how dropping low variance features gives no downside in the proposed work. This research will fill this gap. The paper [26] proposed dictionary-based approach for classifying sentences using supervised learning techniques. Finally, the sentiments were analyzed and after identifying the semantic priority, the summary is generated. The mathematical models were discussed in detail for computing polarity using SentiWordNet. The comparison is presented by taking different datasets.

6.2.1 Supervised Learning Techniques

The supervised learning techniques take the labeled dataset (here words, sentences, or documents) as input, find the patterns and important features inside the dataset, train the model according to these set of important features, and classify the dataset into pre-specified labeled classes. Testing is another phase where validation of the learning model is done. These learning techniques are more controlled data partitioning techniques than unsupervised learning techniques. There is a supervised guidance provided to the training algorithm, also called teacher's information, to move the training in a specific direction. Mathematically, the labeled class y_i can be predicted for an instance x_i as a function $f(\cdot)$ as given below:

$$Y_i = f(x_i) \tag{6.1}$$

In supervised learning, the function $f(\cdot)$ is a set of discrete values, for example, in opinion mining set of values like {fake, not fake}. These sets of discrete values can be taken from unordered and discrete set of values $\{1, 2, \ldots n\}$, each value represents a labeled class [27]. In most of the cases, classification is binary as it is convenient and less complex to classify the data instance in two classes. To reduce the complexity, multi-label classification can be converted to multiple binary classifications. If y_i is taken as the dependent variable of independent variable x_i and represented as numerical values, then this problem can be referred to as regression problem. This section will provide the brief overview of classification algorithms that are used in modeling the summarization techniques.

In text summarization, binary Bayesian classifiers are the oldest approach used for classification [28, 29]. The authors, Aone et al., assigned features to each sentence and calculated the probability of every sentence and included in summary using Bayes' rule. The Naïve Bayes techniques include the assumption of feature dependency, whereas hidden Markov model (HMM) along with logistic regression used by the authors of [30] summarizes the documents by joint distribution of collecting features. Burges et al. used gradient-decent-based neural network to find the ranks of the summarizing sentences, also known as RankNet [31]. Extending the functionality of RankNet, NetSum was proposed in [32] in 2007. Svore et al. have used Lambda Rank to score the sentences and then picked up the highest rank sentences. Further trained the two-layer neural network with Lambda Rank algorithm and used it to summarize the documents from the third-party sources. A more sophisticated semi-supervised learning machine known as the support vector machine has been used by the authors [33] for text summarization based on query or features. In the following year, an ensemble of machine learning technique has been proposed by Xiaojun Wan to do Chinese sentiment analysis using English sentiment resources [34]. In 2009, a structural SVM was used by Li et al. who considered the coverage and diversity for document summarization [35]. An ensemble of feed-forward NN, Gaussian mixture model, genetic algorithm, and mathematical regression were used for feature training in 2009 [36]. In 2014, authors of [37] summarized the single lingual document by using mimetic algorithm and outperformed the many state-of-the-art techniques for summarization. In 2017, Belkebir and Guessoum have used supervised AdaBoost for Arabic text summarization [38]. An automatic text summarization has proposed by the authors in 2018, and corpus was used to find relevant feature to rank the document and then a Naive Bayes classifier to the best score sentence [39].

6.2.2 Unsupervised Learning Techniques

Unlike the supervised learning, unsupervised learning is less controlled learning due to the absence of teaching guidance. Conroy and O'leary [40] have used hidden Markov model (HMM). They calculated the Markovity, a measure based on statistical features for inclusion or exclusion of the summarizing sentence in summary. Fung et al. [41] have proposed unsupervised multi-document summarization technique which combines modifies K-means vector space clustering with the heuristic probabilistic framework. It classifies the major articles by HMM transitional probabilities and tags the sentences and paragraph using segmental K-means algorithm. A semi-supervised approach of using SVM as supervised learning and probabilistic SVM along with Naïve Bayes' classifiers for unsupervised learning which exploits the unlabeled data has developed by Wong et al. [42]. Lobo and Deshpande [43] in 2013 used another clustering technique for summarization which combined the sentences and documents clustering. It calculates the scores of sentences based on features and documents that are clustered using cosine similarity. Further, clustering is done for each document, and best sentences are clustered. They developed the query-based clustering approach which suffers difficulty in sentence clustering. Dohare et al. [44] have developed Semantic Abstractive Summarization which generates Abstract Meaning Representation graph of inputs and then clusters the sentences based on this AMR graph. A number of supervised and unsupervised approaches have been used by Mao et al. [45] to measure the sentence score extracted from a single document. Then, they have compared (i) supervised and graph model separately, (ii) combination of supervised feature extraction and graph model, and last (iii) supervised model and biased graph model on different datasets. They concluded the third method as the most promising combination to summarize the documents. Karaa et al. have combined genetic algorithm with an agglomerative algorithm and vector space model to apply unsupervised learning on the MEDLINE biomedical database for summarization and information retrieval [46]. It is an application of unsupervised clustering into medical and health care. In recent years, many other researchers have retrieved the information and summarize the healthcare data and improved the decision [29, 47, 48].

6.3 Summary Generation

The human's tendency is to formulate a summary which is achieved by determining which opinions are important and which are not. Both approaches initially accept the text document in the form of tokens. The steps are shown in Fig. 6.5 for opinion summarization.

6.3.1 Opinion Aggregation

The proposed work is inspired from the abstractive summarization framework as discussed by authors of [49]. The algorithm starts with the construction of graphs from text and exploring the properties of graphs such as scoring its several sub-paths, ensuring validity and removing redundancy to finally generate candidate summaries. This differs from Ganesan methodology as it fuses sentiment by employing the classification task achieved by CNN and allowing the user to view summarization results taking aspects into account instead of using SentiWordNet. Different researchers have used various techniques by exploiting graphs [18, 50].

A novel approach has been proposed that generates the extractive summary from the set of reviews. It aims to reduce the number of dimensions by dropping irrelevant thematic words and relatively finding the summary of the reviews by taking the

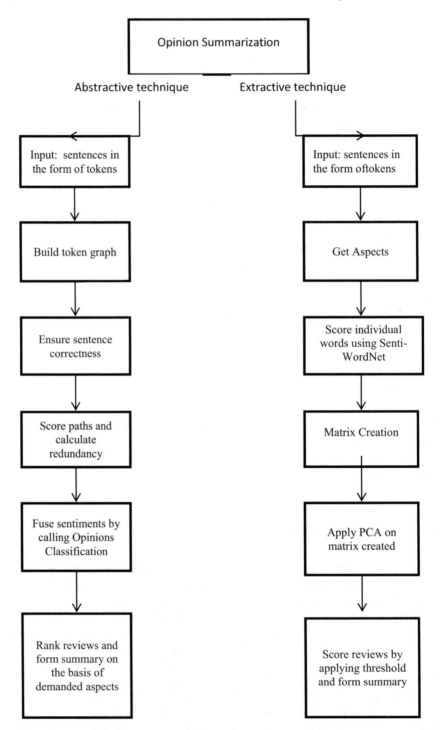

Fig. 6.5 Proposed steps for summarization

reviews according to their rank and ranking is done according to the prime aspects without any loss of information, respective to a particular domain. This method will safely remove those aspects which are not in the top priority. Instead of traversing and searching the reviews on different sites, the decision can be taken by the user quickly by looking at the summary of the corpus. Vector representation of sentences in the summary is best approximated by projecting them into the boundary of the principal components.

6.4 Aspect-Based Opinion Summarization

Abstractive summarization and extractive summarization are accomplished using two different techniques, graph-based method and PCA method, respectively. The analysis conducted on different datasets and comparison made between both of the techniques to discuss which method generates more coherent and complete summary is discussed in the next chapter. The evaluation is done on the basis of ROUGE tool by calculating ROUGE scores. The summaries are produced on sample Opinosis dataset containing 'food' reviews and real dataset consisting 'iPhone' reviews. The accuracy results are reported with F-scores. The algorithm for the above-explained two techniques is presented in detail in the next section.

Algorithm for Abstractive Summarization

Input: I = opinions with explicitly mentioned aspect
Output: O = summaries.

The following are the steps for generating summary:

Step 1: Constructing Graphs
Since the two basic components of graphs are nodes and edges, this approach relies on constructing graphs where each node is stored which represents a word in the text and the edges represent the adjacency of the words in the sentence. The graphs constructed which captures redundancy is shown in Fig. 6.6.

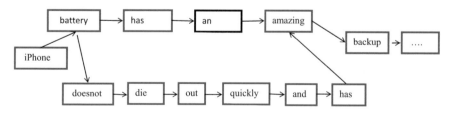

Fig. 6.6 Construction of graph

Every node is a correct node represented as N (tag, p_t, p_s), where tag is the information about the POS tag of the word in that node, p_t is the position of the word in the sentence, and p_s is the point where it is found in the document.

Step 2: Ensure the Accuracy of the Sentence

In the sentence 'iPhone battery has an amazing backup' and 'iPhone battery does not die out quickly and has amazing backup' contain the main aspect 'battery.' All the paths constructed using graphs are correct if they satisfy the following set of rules:

- All the nodes are linked together with directed edges, correct start node (CSN) ensures it is the first node and correct end node (CEN) ensures the sentences are completed.
- The accuracy of the sentence is ensured by the set of rules explained as under:

> /nn following /vb and an /jj
> /jj following /nn and a /vb
> /vb following /jj and a /nn
> /rb following /jj and a /nn
> /rb following /nn

Where /nn is a noun, /vb is a verb, /jj is an adjective, /vb is a verb, and /rb is an adverb. Conjunction can appear in between of a sentence or at the start of a sentence.

Step 3: Scoring Sentences and Merging Sentiments

After ensuring the accuracy of the sentence, the paths are scored by calculating overlap. The redundancy is calculated using the intersection of the position of the words in the sentences such that the difference between the positions is no greater than a threshold, P. The nodes are checked and if they can be fused, the sentiments of those sentences are calculated using opinion classification. The connectors are selected accordingly from the stored list of connectors. For example, in the above Fig. 6.6, two sentences are taken, 'iPhone battery has an amazing backup' and 'iPhone battery does not die out quickly.' These two sentences are connected via connector 'and,' referring to the common aspect 'battery.' Individually, the sentiments of two sentences are calculated as both can be fused and the connector 'and' is used as it is connected both the sentences accurately. The whole graph is traversed again to find further nodes. The new score is computed and the duplicate sentences are removed from the fused sentences. The resultant fused sentences are added to their scores.

Step 4: Ranking of sentences for summarization

The rest of the sentences are sorted in descending order and top n sentences are chosen to be the candidate summary (n is chosen by the user).

Principal Component Analysis (PCA)

Karl Pearson (1901) invented PCA. It is a statistical technique and dimension reduction tool, i.e., used to decrease the dimension of the dataset and reduce the correlation between variables. PCA identifies various patterns present in the data and representing the data in such a manner that their similarities and dissimilarities are highlighted. The purpose of incorporating it for summarization is to safely eliminate many features from the data space [51]. Smith [52], in 2002, discovered PCA useful in many areas such as face recognition and image compression. The major advantage of using PCA is after finding the patterns in the data, and the compression task becomes easier. Thus, dimensions can be reduced without losing important information [53]. The proposed work is to incorporate PCA in the text so that the most important relevant reviews can be extracted from the list of reviews. This technique is purely novel and the analysis of the results after comparing the peer summaries with the gold summaries will prove that the proposed technique is unique and the results are more accurate.

Algorithm for Extractive Summarization
A novel approach has been proposed that generate the extractive summary from the set of reviews. It aims to determine all relevant sentences and reducing dimensions by dropping irrelevant thematic words and non-essential sentences by applying combination of PCA and SVD.

The following steps are involved in generating summary:
Input: I = opinions with explicitly mentioned aspect
Output: O = summaries

Step 1: Score Individual Words Using SentiWordNet
For each identified thematic-opinion word pair by using aspect identification, score from the database (SentiWordNet) is retrieved.

Step 2: Matrix Creation for Summarization
There are numbers of thematic Words. Each nth thematic word will be associated with m number of reviews.
Matrix T created as follows: $T = n * m$
Values of matrix $T[ij]$ will represent mth opinion word of nth thematic word.

Step 3: Implementing PCA: Applying PCA on Matrix Created
Apply SVD to the created matrix. The result of SVD will be three matrices S, U, VT, where S, U, VT represent matrix. The principal components of any matrix are its eigenvalues.

U and V matrix will be considered up to rank K (i.e., ignoring lower sparse part of those matrices) called as rank-k approximation. The value of k is a parameter whose choice is of critical importance. The counter is applied for each thematic word to get the value of its frequency count, i.e., term frequency of each thematic word. The thematic word with the highest counter value is taken as a query vector. Cosine similarities, i.e., measure of similarity between the query vector and review vector is calculated using the below formula

$$(q,r) = q \cdot r / |q| \cdot |r| \qquad\qquad (6.2)$$

where q is the reduced query vector and r is the review vector.

Step 4
After sorting the values of cosine similarity in descending order, the threshold is applied to the sorted values.

6.5 Summary

Structured feature-based summarization of opinions with relevant information has been formed and presented to the users. Opinion summarization using graphs helps in analyzing the sentiment in the review by considering the semantic knowledge. Thus, the main focus of this research is to propose a semantic framework in all stages of feature-based opinion summarization. Thus, explicit and implicit features in the reviews and its sentiment value are expressed for each feature quantitatively and an integrated method is proposed which defines our purpose of the structured opinion summarization. The final summarized opinion is presented according to the query in respective domainLarge corpus of online posted reviews that has been summarized using two different methodologies. The proposed abstractive technique gives summarized sentences in a beautifully structured way including the relevant and important segments of the sentence. The algorithm is proposed by constructing graphs as a preliminary task that denotes the richness of representational structure of sentences. Results show that extractive summarization does not require much hard work in understanding the content of opinions or understanding text in depth. It can be easily achieved by ranking and by choosing threshold that will filter out the most appropriate sentences to be included in the summary. Further, manual evalua-tion by human judges shows that the proposed technique produces summaries with acceptable linguistic quality and high in formativeness.

References

1. Bhatia, S. (2014, April). New improved technique for initial cluster centers of K means clus-tering using Genetic Algorithm. In *International Conference for Convergence for Technology-2014* (pp. 1–4). IEEE.
2. Bhatia, S. (2012). Breaking AES using genetic algorithm. In *Proceedings of 4th International Conference on Electronics Computer Technology (ICECT)* (p. 6), Kanyakumari, India. IEEE. ISBN: 978-1-46731849-5.
3. Liu, C. L., Hsaio, W. H., Lee, C. H., Lu, G. C., & Jou, E. (2012a). Movie rating and review summarization in mobile environment. *IEEE Transactions on Systems, Man, and Cybernetics, Part C (Applications and Reviews), 42*(3), 397–407.

4. Liu, J., Seneff, S., & Zue, V. (2012b). Harvesting and summarizing user-generated content for advanced speech-based HCI. *IEEE Journal of Selected Topics in Signal Processing, 6*(8), 982–992.

5. Raut, V. B., & Londhe, D. D. (2014). Survey on opinion mining and summarization of user reviews on web. *International Journal of Computer Science and Information Technologies, 5*(2), 1026–1030.

6. Tayal, M. A., Raghuwanshi, M. M., & Malik, L. G. (2017). ATSSC: Development of an approach based on soft computing for text summarization. *Computer Speech & Language, 41,* 214–235.

7. Himabindu, K., Morusupalli, R., Dey, N., & Rao, C. R. (2019). Coefficient of variation and machine learning applications. CRC Press.

8. Singh, A., Dey, N., & Ashour, A. S. (2017). Scope of automation in semantics-driven multimedia information retrieval from web. In *Web Semantics for Textual and Visual Information Retrieval* (pp. 1–16). IGI Global.

9. Steinberger, J., Poesio, M., Kabadjov, M. A., & Ježek, K. (2007). Two uses of anaphora resolution in summarization. *Information Processing and Management, 43*(6), 1663–1680.

10. Lan, K., Wang, D. T., Fong, S., Liu, L. S., Wong, K. K., & Dey, N. (2018). A survey of data mining and deep learning in bioinformatics. *Journal of Medical Systems, 42*(8), 139.

11. Gupta, V., & Lehal, G. S. (2010). A survey of text summarization extractive techniques. *Journal of Emerging Technologies in Web Intelligence, 2*(3), 258–268.

12. Silber, H. G., & McCoy, K. F. (2000, January). Efficient text summarization using lexical chains. In *Proceedings of the 5th international conference on Intelligent user interfaces* (pp. 252–255). ACM.

13. Carenini, G., & Cheung, J. C. K. (2008, June). Extractive vs. NLG-based abstractive summarization of evaluative text: The effect of corpus controversiality. In *Proceedings of the Fifth International Natural Language Generation Conference* (pp. 33–41). Association for Computational Linguistics.

14. Genest, P. E., & Lapalme, G. (2012, July). Fully abstractive approach to guided summarization. In *Proceedings of the 50th Annual Meeting of the Association for Computational Linguistics: Short Papers-Volume 2* (pp. 354–358). Association for Computational Linguistics.

15. Clarke, J., & Lapata, M. (2006, July). Models for sentence compression: A comparison across domains, training requirements and evaluation measures. In *Proceedings of the 21st International Conference on Computational Linguistics and the 44th annual meeting of the Association for Computational Linguistics* (pp. 377–384). Association for Computational Linguistics.

16. Knight, K., & Marcu, D. (2002). Summarization beyond sentence extraction: A probabilistic approach to sentence compression. *Artificial Intelligence, 139*(1), 91–107.

17. Zajic, D., Dorr, B. J., Lin, J., & Schwartz, R. (2007). Multi-candidate reduction: Sentence compression as a tool for document summarization tasks. *Information Processing and Management, 43*(6), 1549–1570.

18. Liu, F., Flanigan, J., Thomson, S., Sadeh, N., & Smith, N. A. (2015). Toward abstractive summarization using semantic representations.

19. Wang, D., Li, T., Zhu, S., & Ding, C. (2008, July). Multi-document summarization via sentence-level semantic analysis and symmetric matrix factorization. In *Proceedings of the 31st annual international ACM SIGIR conference on Research and development in information retrieval* (pp. 307–314). ACM.

20. Sankar, K., & Sobha, L. (2009, June). An approach to text summarization. In *Proceedings of the Third International Workshop on Cross Lingual Information Access: Addressing the Information Need of Multilingual Societies* (pp. 53–60). Association for Computational Linguistics.

21. Ganesan, K., Zhai, C., & Han, J. (2010, August). Opinosis: a graph-based approach to abstractive summarization of highly redundant opinions. In *Proceedings of the 23rd international conference on computational linguistics* (pp. 340–348). Association for Computational Linguistics.

22. Karaa, W. B. A., & Dey, N. (2017). *Mining multimedia documents.* Chapman and Hall/CRC.

23. Dahiya, K., & Bhatia, S. (2015, September). Customer churn analysis in telecom industry. In *2015 4th International Conference on Reliability, Infocom Technologies and Optimization (ICRITO) (Trends and Future Directions)* (pp. 1–6). IEEE.
24. Bhargava, R., Sharma, Y., & Sharma, G. (2016). ATSSI: Abstractive text summarization using Sentiment infusion. *Procedia Computer Science, 89,* 404–411.
25. Gawalt, B., Zhang, Y., & El Ghaoui, L. (2010). Sparse pca for text corpus summarization and exploration. In *NIPS 2010 Workshop on Low-Rank Matrix Approximation*.
26. Khan, F. H., Qamar, U., & Bashir, S. (2016). A semi-supervised approach to sentiment analysis using revised sentiment strength based on SentiWordNet. *Knowledge and Information Systems,* 1–22.
27. Kamal, M. S., Dey, N., & Ashour, A. S. (2017). Large scale medical data mining for accurate diagnosis: A blueprint. In *Handbook of large-scale distributed Computing in smart healthcare* (pp. 157–176). Cham: Springer.
28. Aone, C., Okurowski, M. E., & Gorlinsky, J. (1998, August). Trainable, scalable summarization using robust NLP and machine learning. In *Proceedings of the 17th international conference on Computational linguistics-Volume 1* (pp. 62–66). Association for Computational Linguistics.
29. Fong, S., Li, J., Song, W., Tian, Y., Wong, R. K., & Dey, N. (2018). Predicting unusual energy consumption events from smart home sensor network by data stream mining with misclassified recall. *Journal of Ambient Intelligence and Humanized Computing, 9*(4), 1197–1221.
30. Schlesinger, J. D., Okurowski, M. E., Conroy, J. M., O'Leary, D. P., Taylor, A., Hobbs, J., & Wilson, H. T. (2002). Understanding machine performance in the context of human performance for multi-document summarization.
31. Burges, C., Shaked, T., Renshaw, E., Lazier, A., Deeds, M., Hamilton, N., & Hullender, G. N. (2005). Learning to rank using gradient descent. In *Proceedings of the 22nd International Conference on Machine learning (ICML-05)* (pp. 89–96).
32. Svore, K., Vanderwende, L., & Burges, C. (2007, June). Enhancing single-document summarization by combining RankNet and third-party sources. In *Proceedings of the 2007 Joint Conference on Empirical Methods in Natural Language Processing and Computational Natural Language Learning (EMNLP-CoNLL)* (pp. 448–457).
33. Li, S., Ouyang, Y., Wang, W., & Sun, B. (2007). Multi-document summarization using support vector regression. In *Proceedings of DUC.*
34. Wan, X. (2008, October). Using bilingual knowledge and ensemble techniques for unsupervised Chinese sentiment analysis. In *Proceedings of the Conference on Empirical Methods in Natural Language Processing* (pp. 553–561). Association for Computational Linguistics.
35. Li, L., Zhou, K., Xue, G. R., Zha, H., & Yu, Y. (2009, April). Enhancing diversity, coverage and balance for summarization through structure learning. In *Proceedings of the 18th International Conference on World Wide Web* (pp. 71–80). ACM.
36. Wan, X. (2008). Using only cross-document relationships for both generic and topic-focused multi-document summarizations. *Information Retrieval, 11*(1), 25–49.
37. Mendoza, M., Bonilla, S., Noguera, C., Cobos, C., & León, E. (2014). Extractive single-document summarization based on genetic operators and guided local search. *Expert Systems with Applications, 41*(9), 4158–4169.
38. Belkebir, R., & Guessoum, A. (2015). A supervised approach to Arabic text summarization using adaboost. In *New Contributions in information systems and technologies* (pp. 227–236). Cham: Springer.
39. Kumar, V., Choudhury, T., Sabitha, A.S., & Mishra, S. (2019). Summarization using corpus training and machine learning. In Emerging *Trends in expert applications and security* (pp. 555–563). Singapore: Springer.
40. Conroy, J. M., & O'leary, D. P. (2001, September). Text summarization via hidden markov models. In *Proceedings of the 24th Annual International ACM SIGIR Conference on Research and Development in Information Retrieval* (pp. 406–407). ACM.
41. Fung, P., Ngai, G., & Cheung, C. S. (2003, July). Combining optimal clustering and hidden Markov models for extractive summarization. In *Proceedings of the ACL 2003 Workshop on Multilingual Summarization and Question Answering-Volume 12* (pp. 21–28). Association for Computational Linguistics.

42. Wong, K. F., Wu, M., & Li, W. (2008, August). Extractive summarization using supervised and semi-supervised learning. In *Proceedings of the 22nd International Conference on Computational Linguistics-Volume 1* (pp. 985–992). Association for Computational Linguistics.

43. Deshpande, A. R., & Lobo, L. M. R. J. (2013). Text summarization using Clustering technique. *International Journal of Engineering Trends and Technology, 4*(8), 3348–3351.

44. Dohare, S., Gupta, V., & Karnick, H. (2018, July). Unsupervised semantic abstractive summarization. In *Proceedings of ACL 2018, Student Research Workshop* (pp. 74–83).

45. Mao, X., Yang, H., Huang, S., Liu, Y., & Li, R. (2019). Extractive summarization using supervised and unsupervised learning. *Expert Systems with Applications, 133,* 173–181.

46. Karaa, W. B. A., Ashour, A. S., Sassi, D. B., Roy, P., Kausar, N., & Dey, N. (2016). Medline text mining: An enhancement genetic algorithm based approach for document clustering. In *Applications of Intelligent Optimization in Biology and Medicine* (pp. 267–287). Cham: Springer.

47. Varlamis, I., Apostolakis, I., Sifaki-Pistolla, D., Dey, N., Georgoulias, V., & Lionis, C. (2017). Application of data mining techniques and data analysis methods to measure cancer morbidity and mortality data in a regional cancer registry: The case of the island of Crete, Greece. *Computer Methods and Programs in Biomedicine, 145,* 73–83.

48. Karaa, W. B. A. (Ed.). (2015). *Biomedical image analysis and mining techniques for improved health outcomes.* IGI Global.

49. Ganesan, K. A. (2013). *Opinion driven decision support system.* University of Illinois at Urbana-Champaign.

50. Kumar, N., Srinathan, K., & Varma, V. (2013). A knowledge induced graph-theoretical model for Extract and abstract single Document summarization. *Computational linguistics and intelligent text processing* (pp. 408–423). Berlin Heidelberg: Springer.

51. Hofmann, T. (1999, August). Probabilistic latent semantic indexing. In *Proceedings of the 22nd Annual International ACM SIGIR Conference on Research and Development in Information Retrieval* (pp. 50–57). ACM.

52. Smith, L. I. (2002). A tutorial on principal components analysis. *Cornell University, USA, 51*(52), 65.

53. Vinodhini, G., & Chandrasekaran, R. M. (2014). Opinion mining using principal component analysis based ensemble model for e-commerce application. *CSI Transactions on ICT, 2*(3), 169–179.

Chapter 7
Conclusions

This book focuses on developing an overall system for summarizing opinions based on aspects. There are mainly three pieces of work:

The first piece of work discusses how to extract opinions from social media and how to perform preprocessing tasks. Social media serve as a global source of posting feedbacks from different people on various products. So, on considering some seed URLs, Chap. 3 discusses extraction of opinions in a timely fashion such that the opinions which are recently posted can also be extracted by calculating the precedence of the seed URLs. The preprocessing tasks are performed on the crawled reviews in order to clean and prepare further for summarizing opinions.

The second task is focused on two different modules: extraction of aspects and classification of opinions. Aspect identification used a particular rule-based approach to find syntactical dependency relations, focused on aspects, and their evaluations to collect target opinion-word pairs. Opinion classification used the algorithms of deep learning by presenting CNN model for classifying text with the least intervention of human. The one-layer architecture of CNN is used in which opinions in the form of tokens were converted into vectors using Word2Vec embedding and three basic layers (convolution layer, pooling layer, and softmax layer) were employed for classifying opinions into two classes, i.e., positive and negative. The feature-level sentiment classification is achieved by classifying the opinions into positive and negative categories by specifying the aspect, the user is interested in viewing.

In the next step, summarization task is discussed. Since the summary is to be done on the basis of aspects, the algorithm is proposed which can provide structured well-formed summary merging important parts of sentences. Two different approaches of summarization are discussed: abstractive and extractive. The abstractive method relies on constructing graphs, ensuring the accuracy of the sentence, calculating redundancy, and linking of nodes using sentiment scores achieved by previously explained opinion classification module. Extractive-based technique solely depends on PCA method based on the statistical method. It works by finding eigenvalues using SVD, setting priority of the sentence by calculating rank of individual sentences based

© The Author(s), under exclusive license to Springer Nature Singapore Pte Ltd. 2020
S. Bhatia et al., *Opinion Mining in Information Retrieval*,
SpringerBriefs in Computational Intelligence,
https://doi.org/10.1007/978-981-15-5043-0_7

on threshold and sorting them in descending order to filter the important sentences to be included in the summarized list.

Finally, all the above three tasks are identified, starting from crawling of opinions using opinion retriever, to extraction of aspects using grammar dependency rules, classification of sentences using CNN model and presenting summarized opinions based on the aspects.

7.1 Tools and Techniques

As the documents contain the textual data and images, it is difficult to extract the information like emotions and opinions. Another issue is the language of communication and structure of the data from heterogeneous sources. So, there is a wide range of tools available in the domain. They are ready to use tools which automate the pre-processing, feature extraction, feature selection, classification, clustering, etc. Due to easy to use interface environment and availability of in-built libraries, WEKA and NLTK (Python Framework) are the most widely used tools. Other examples include CoreNLP, Apache openNLP, Gate, Pattern another Python framework, LingPipe, and opinion finder. Here, in this section, we are discussing Natural Language Tool Kit (NLTK), a Python Framework which includes many in-built libraries to ease the opinion mining task.

It provides a complete package of text processing libraries, NLP libraries, parsing, tokenization, tagging, classification, stemming, and semantic reasoning, and an active discussion forum along with user-friendly libraries to many corpora and lexical resources like WordNet. Graphical demonstrations and sampling of data using NLTK are straightforward. NLTK is projected to support teaching, analyzing, and research in NLP or closely related areas, including machine learning, empirical linguistics, information retrieval, artificial intelligence, and cognitive science. It includes the tokenization to split a string into subclasses and handle the operators and delimiters. Further stemming can be done using NLTK stem, an interface in NLTK. Stemming is a part of normalizing method. Using NLTK, further steps of segmentation, collation [1], tagging [2], followed by parsing [3] can be accomplished. The multi-purpose NLTK can be used successfully as an individual study tool, a teaching tool, or as a platform for prototyping and building research systems.

Other tools like Stanford CoreNLP use bootstrapped pattern learning, the POS tagging, parsing, coreference resolution system, named-entity recognizer, and sentiment analysis. Apache OpenNLP uses named-entity extraction, tokenization, part-of-speech tagging, chunking, parsing, and coreference resolution. LingPipe uses POS tagging, classification, entity extraction, and clustering. GATE uses tokenizer, sentence splitter, named entities transducer, POS tagging, gazetteer, and coreference tagger. Pattern, a Python Framework, uses WordNet, data mining, N-gram search, sentiment analysis, POS tagging, visualization, machine learning, and network analysis. Weka is another easy-to-use tool which is discussed in the following section of the chapter.

There are various tools available for opinion mining. Some of them are as follows:

Review Seer helps in automation of the work through aggregation sites. The extracted feature terms are assigned a score using NB classifier by collecting a set of positive and negative opinions. The output is in the form of review sentences with the score of each feature [4, 5].

Web Fountain makes use of definite Base Noun Phrase (dBNP) by extracting the product features following a heuristic approach. Two resources which are used in this tool are sentiment lexicon and sentiment pattern database. Opinion sentences are parsed, and sentiments are provided to each feature with this tool [6].

Red Opal helps the user to determine the entities and its polarity of the sentence is found on the basis of aspects. The orientation of the reviews is found on the basis of attributes from the opinions. The output is shown on a Web-based interface [7, 8].

Opinion Observer is a sentiment analysis tool which displays output graphically by presenting a comparison of the analysis made from the opinions posted online. It uses WordNet exploring method to assign prior polarity.

The following tools are used to find polarity from text:

Waikato Environment for Knowledge Analysis (Weka) is a java suite of machine learning, used to compare the various machine learning algorithms on the basis of accuracy, precision, and recall for data mining applications [9, 10].

RapidMiner helps in giving solutions and services in the area of analytics, data mining, and text mining applications.

Recall-Oriented Understudy for Gisting Evaluation (ROUGE) the data summarization tool, stated by Lin in 2004 [11]. It is used to effectively calculate the worth of summary generated between humans and systems. The overlapping units are counted among both summaries such as the N-gram, word sequences, and word pairs.

NLP Tools General Architecture for Text Engineering (GATE) is NLP and language engineering tool. Natural Language Toolkit (NLTK), LingPipe, OpenNLP, Stanford Parser, and POS tagger, etc., Kaur and Saini [12], are part of NLP-based tools.

7.2 Datasets and Evaluations

Opinion mining lies at the crossroads of computational linguistics and IE. Therefore, the proposed work combines theories of supervised learning techniques with NLP tasks in order to develop the complete IR system. The datasets used in this research are listed in Table 7.1.

The book covers all the concepts of developing an IR opinion mining system with the following objectives:

- The existing Web search engine is used to develop an opinion retriever for extracting opinions. The novel algorithms are proposed in order to detect Web pages by

Table 7.1 Dataset used for the research

Dataset	Type	Domain
Multi-domain sentiment dataset (MDS) [13]	Text	Products (books, DVDs, electronics, kitchen appliances)
Movie review dataset (MR) [14]	Text	Movies
Opinosis dataset [15]	Text	Hotels, cars, products

calculating revisit time to gather fresh opinions. The extraction and ranking of objects of interest are done after applying preprocessing tasks.

- The aspects are identified by proposing rules based on grammar dependency. After detecting aspects using NLP tools and dependency relations (extracting opinion word feature pair like amazing battery, pretty ugly voice quality, etc.), sentiment analysis is performed.
- The sentences are categorized into one of the two categories positive and negative based on the learning algorithms of deep learning. Classification of opinions is done by selecting appropriate parameters and configurations using CNN.
- The abstractive summarization of sentences is achieved by exploiting the graph structure of a sentence using NLP approaches. The extractive summary is formed using PCA. Adequately, comparison with both the techniques is performed and results of the same are analyzed and compared using Recall-Oriented Understudy for Gisting Evaluation (ROUGE) tool.

Evaluation:

The opinions are extracted from seed URLs and preprocessing operations are applied to the opinions. Grammatical relations are explored which are used to find the relationship between target entities and features on the opinions. Examples of opinion words are poor, beautiful, great, etc., and target entities are battery, picture, speed, etc. Then, the semantic orientations of opinions are determined based on the aspects identified. Sentiment classification at the aspect level is the next prime task, which classifies each sentence based on the aspects into positive and negative categories. The algorithm of deep learning, i.e., CNN, is applied to the opinions for categorizing opinions. Finally, ranking and summarization of opinions are achieved using PCA technique and abstractive summary is generated using graphs.

The following set of seed URLs has been taken for the experimental evaluation of this work which is given in Chap. 3. The query taken from electronics domain is listed as 'iPhone 7s'.

For evaluating the work, near about 55–65 response pages corresponding to each URL are collected for the dataset iPhone 7s. In total, a sample of 200–250 pages has been gathered with a total of 1797 opinions. The analysis is shown in Table 7.2.

Thus, the accuracy of the proposed work is much high in terms of precision and recall. The analysis of the same has been achieved which is shown in the graph in Fig. 7.1 (Table 7.3).

The crawling task is based on the corpus study, and results are evaluated in terms of precision, recall, and F-measure.

Table 7.2 Seed URLs

S. No.	Seed URLs
1	www.amazon.com
2	www.flipkart.com
3	www.yelpme.com
4	www.sitejabber.com

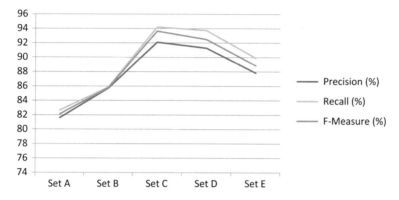

Fig. 7.1 Performance analysis

Table 7.3 Performance values

No. of pages	Precision (%)	Recall (%)	F-Measure (%)
Set A	81.61	82.66	82.13
Set B	85.70	85.86	85.77
Set C	92.10	94.20	93.64
Set D	91.27	93.73	92.48
Set E	87.82	89.90	88.84

For evaluation of the proposed opinion summarization work in Chap. 6, the comparison is made between the abstractive technique and the extractive technique for Opinosis dataset and the dataset explained above. It was found that extractive methods have very high recall scores as compared to precision scores. The criteria of 'selecting sentences' and not much significant information in summary results in giving higher values for recall but low values for precision, respectively. Tables 7.4 and 7.5 show the results.

The comparison is done for Opinosis dataset [15] and found that better results are reported with the proposed algorithm. The results are shown in Figs. 7.2, 7.3 and 7.4.

The above results show that the proposed technique has comparable results with human summaries. The higher precision is reported because of the limitation as discussed previously in the paper [15], that it can connect only those sentences where prior connector existed for fusing sentences. The proposed technique outperforms by

Table 7.4 Results with Opinosis dataset

	Abstractive-based summarization		Extractive-based summarization	
	Rouge 1	Rouge 2	Rouge 1	Rouge 2
Precision	0.4983	0.2529	0.0826	0.0175
Recall	0.2719	0.0791	0.5045	0.1106
F-Measure	0.3518	0.1205	0.1419	0.0302

Table 7.5 Results with Iphone dataset

	Abstractive-based summarization		Extractive-based summarization	
	Rouge 1	Rouge 2	Rouge 1	Rouge 2
Precision	0.46	0.32	0.14	0.07
Recall	0.19	0.15	0.62	0.13
F-Measure	0.26	0.20	0.22	0.09

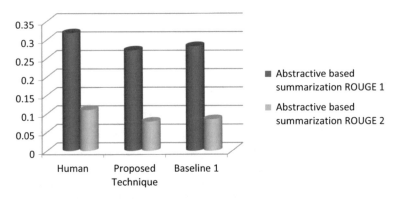

Fig. 7.2 Results with recall for abstractive summarization on Opinosis dataset

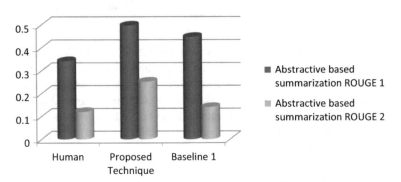

Fig. 7.3 Results with precision for abstractive summarization on Opinosis dataset

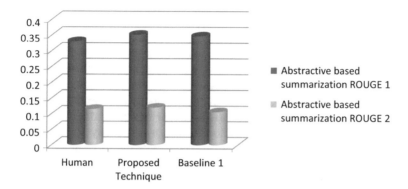

Fig. 7.4 Results with F-measure for abstractive summarization on Opinosis dataset

a good margin. Low results on recall are because of immense existence of duplicate information on dataset.

7.3   Future Work

Throughout the context of the research, the different algorithms and approaches are discussed to develop an IR system for mining opinions. While listed objectives in Chap. 1 are accomplished to develop a summary of opinions, there were some points which are retained for the future work:

- Big data opens up huge scope for 'opinion mining' and 'sentiment analysis'. Since this work is limited to a small amount of data, big data can give an opportunity to look upon huge volumes of data. Big data will be able to perform the task of classification in a better way overcoming its challenges.
- Since the proposed work is based on the application of NLP that uses syntactical rules for fining thematic-opinion word, novel algorithms for identifying aspects, classifying sentiments and summarizing opinions are proposed, incorporating changes in these rules, and algorithms can make this system compatible with other languages such as Hindi and Arabic.
- Tweets, reviews, blogs/chats, and other review sites also consider ill-formatted text which may have a lot of spelling errors, missing punctuation symbols, and other mistakes as well. Although, in this research work, this issue has been looked upon at moderate level, still incorporating removal techniques for this kind of text will make the proposed system more durable and robust.
- Some other algorithms of deep learning can also be applied such as autoencoders and long short-term memory (LSTM). These networks consider training and mapping which automatically classify the new input vector. These can be studied further for finding out the effectiveness of the research work.

7.4 Summary

The three major research questions that lie in this work are 'how the opinions are extracted?', 'what approach is carried for classification?' and 'how the structured aspect based summary is generated?' These three research questions put forward the basis to provide a solution to the problem of computationally forming a complete IR system, and their reasoning is discussed in the book. This book addresses the problem of how to classify sentiments and develop the opinion system by combining theories of NLP and deep learning. Comparative experiments on different datasets which contain real discussions and reviews are conducted and the accuracy is effectively measured using performance metrics and data mining tools.

References

1. Navigli, R., & Velardi, P. (2004). Learning domain ontologies from document warehouses and dedicated web sites. *Computational Linguistics, 30*(2), 151–179.
2. Huang, E. H., Socher, R., Manning, C. D., & Ng, A. Y. (2012). MaxMax: A graph-based soft clustering algorithm applied to word sense induction. In *Proceedings of the 14th International Conference on Computational Linguistics and Intelligent Text Processing* (pp. 368–381).
3. Moro, A., Raganato, A., & Navigli, R. (2014). Entity linking meets word sense disambiguation: A unified approach. *Transactions of the Association for Computational Linguistics, 2,* 231–244.
4. Maynard, D., Bontcheva, K., & Rout, D. (2012). Challenges in developing opinion mining tools for social media. *Proceedings of the @ NLP can u tag# usergeneratedcontent* (pp. 15–22).
5. Bhatia, M., Sharma, S., Bhatia, S., & Alojail, M., (2020). Fog computing mitigate limitations of cloud computing. *International Journal of Recent Technology and Engineering (IJRTE), SCOPUS*.
6. Gruhl, D., Chavet, L., Gibson, D., Meyer, J., Pattanayak, P., Tomkins, A., & Zien, J. (2004). How to build a WebFountain: An architecture for very large-scale text analytics. *IBM Systems Journal, 43*(1), 64–77.
7. Scaffidi, C., Bierhoff, K., Chang, E., Felker, M., Ng, H., & Jin, C. (2007). Red opal: Product-feature scoring from reviews. In *Proceedings of the 8th ACM Conference on Electronic Commerce* (pp. 182–191). ACM.
8. Bhatia, S., & Madaan, R. (2020). Understanding the role of emotional intelligence in usage of social media. In *Confluence 2020*, Amity University. IEEE.
9. Holmes, G., Donkin, A., & Witten, I. H. (1994). Weka: A machine learning workbench. In *Intelligent Information Systems, 1994. Proceedings of the 1994 Second Australian and New Zealand Conference* (pp. 357–361). IEEE.
10. Chatterjee, S., Datta, B., Sen, S., Dey, N., & Debnath, N. C. (2018, January) Rainfall prediction using hybrid neural network approach. In *2018 2nd International Conference on Recent Advances in Signal Processing, Telecommunications & Computing (SigTelCom)* (pp. 67–72). IEEE.
11. Lin, C. Y. (2004). Rouge: A package for automatic evaluation of summaries. In *Text summarization Branches out: Proceedings of the ACL-04 Workshop* (Vol. 8).
12. Kaur, J., & Saini, J. R. (2015). A study of text classification natural language processing algorithms for Indian languages. *The VNSGU Journal of Science Technology, 4*(1), 162–167.
13. Blitzer, J., Dredze, M., & Pereira, F. (2007). Biographies, bollywood, boom-boxes and blenders: Domain adaptation for sentiment classification. In *ACL* (Vol. 7, pp. 440–447).

14. Pang, B., Lee, L., & Vaithyanathan, S. (2002). Thumbs up? Sentiment classification using machine learning techniques. In *Proceedings of the ACL-02 Conference on Empirical Methods in Natural Language Processing, Association for Computational Linguistics* (Vol. 10 pp. 79–86).
15. Ganesan, K. A., Zhai, C. X., & Han, J. (2010). Opinosis: A graph based approach to abstractive summarization of highly redundant opinions. In *Proceedings of the 23rd International Conference on Computational Linguistics (COLING '10).*

Printed in the United States
By Bookmasters